Oxford Chemistry Series

General Editors
P. W. ATKINS   J. S. E. HOLKER   A. K. HOLLIDAY

Oxford Chemistry Series

E. S. STERN (Editor)
COMPANY PLANNING DEPARTMENT, ICI

The chemist in industry (1):
fine chemicals for polymers

Clarendon Press · Oxford · 1973

Oxford University Press, Ely House, London W.1

GLASGOW NEW YORK TORONTO MELBOURNE WELLINGTON
CAPE TOWN IBADAN NAIROBI DAR ES SALAAM LUSAKA ADDIS ABABA
DELHI BOMBAY CALCUTTA MADRAS KARACHI LAHORE DACCA
KUALA LUMPUR SINGAPORE HONG KONG TOKYO

PRINTED IN GREAT BRITAIN BY
J. W. ARROWSMITH LTD., BRISTOL, ENGLAND

Editor's foreword

TRADITIONALLY, university courses in chemistry provide an academic training but do not generally involve a consideration of modern technological developments. Academic standards of graduates are of vital concern to industry, but it is equally important that chemists in industry should have a sound appreciation of the social and economic aspects of the chemical industry, and of the typical technological problems associated with bringing a new product on to the market.

Fine chemicals for polymers is the first of a number of books in the Oxford Chemistry Series devoted to chemistry and industry. The aim of these books is to provide the necessary background information to enable the graduate entering the chemical industry to develop an appreciation of current industrial development and philosophy. The editor of these books, Dr. E. S. Stern, holds a senior managerial position in Imperial Chemical Industries Ltd. and has been closely associated with many aspects of the modern chemical industry. His key position in this industry ensures that the material covered in the books is expertly treated by a team of authors, each of whom is an acknowledged authority in the area of his contribution.

This book is concerned mainly with specific problems associated with the processing of synthetic yarns and materials, wherever possible on the machinery originally developed for the processing of the traditional natural products, wool and cellulose fibres. The fabrication and dyeing of synthetic fibres pose problems which can be regarded as typical of those to be expected in the development of any new material. The methods by which these and other problems such as product stability have been overcome illustrate the use of 'small tonnage' chemicals as additives to the 'gross tonnage' synthetic polymers.

J.S.E.H.

Preface

CHEMISTRY is fascinating; so are many other academic studies. It is also useful and in this it differs from some other subjects. Because chemistry has these two characteristics, chemists are numerous: about 2000 graduate each year in Britain. They see the fascination of the subject from their academic teachers, and the usefulness of chemistry joined to other academic disciplines—biochemistry, geochemistry, etc. Of its industrial application they know little. Yet, in Britain, only about 10 per cent of all chemists make their career in universities (about 25 per cent in all teaching) while about 55 per cent put chemistry to use in industry.

Few books deal adequately with the role of chemistry in industry, and with its impact on society. There is a need therefore for further information about what the chemist does in industry. This book should provide some understanding of the role and the problems facing the chemist in the industry making and selling fine organic chemicals. The products of this industry serve to upgrade other materials. Their effects are widely used, too widely to be treated comprehensively in one slim book. We have to be selective. We have chosen to discuss those organic 'effects chemicals' used with polymeric materials.

A few words about the background may be helpful. Only those polymers are made which are 'useful', i.e. which someone will buy because they fulfil a need at that price. Large tonnages of polymers are sold—some for use as bulk materials (e.g. polythene as buckets, or elastomers in motor-car tyres), some as film (e.g. polypropylene for packaging), and some as fibres (e.g. in textile uses, so that Nylon became almost synonymous with ladies' stockings). Each material in each suggested or actual use has competitors. Whether one or other of the polymers is in fact used depends on how its performance in use relates to its price, and to the cost and effectiveness of competing materials. Industry constantly tries to reduce the cost of its products and thus to increase their cost-effectiveness.

Another way of improving cost-effectiveness is to add to the products small amounts of chemicals which enhance their performance, or make them acceptable in some other way to the potential user. This forms the subject of this book. Perhaps the most obvious additives making polymers more saleable are colouring matters. Life would be dull indeed if all textile materials were off-white. For this reason dyeing and pigmenting have been practised for centuries. Here we deal with the dyeing of synthetic fibres which in the last 20 years have found increasing use: polyamides (Nylon), polyester, and polyacrylonitrile, the most important three, have all posed problems to the dyestuffs technologist. Other polymers, in bulk material or as film, also have to be coloured. The chemistry involved is intricate and the technology important. The subject provides a good view of the wide scope in industry for the chemist whose interest ranges beyond the academic chemistry.

Less obvious but equally important are other additives: agents conferring stability on the polymer against oxidative or other degradation; cross-linking agents increasing the rigidity or impact resistance; and processing aids permitting material to pass trouble-free through machinery. In each case structure is correlated with activity and cost, and the ultimate choice—often after lengthy testing—falls on the most cost-effective additive.

This book is in two halves. The first deals with stabilizers and pigments, and their chemistry and utility in synthetic polymers. The second part concentrates on synthetic fibres, and the dyes and processing aids for them. Thus some recently-developed applied chemistry is touched on, and technological aspects of fibre-processing are discussed which are unlikely to be found in normal chemistry texts. It is hoped these chapters, going as they do beyond the normal range of academic chemistry, will help the chemist to assess the role he might be called upon to play in industry.

To present this picture of an important part of the fine chemicals industry we are fortunate to have as authors experts equally at home in chemistry and industry. Most sadly, Howard Berrie died suddenly before he had completed his part of the text. A real example of industrial collaboration ensued; his colleagues—anonymously, at their insistence—prepared his chapter for publication. This book is now offered to readers both for their enjoyment and for their education.

E. S. Stern

Contents

† Organics Division Research Department, ICI.

1. Polymer stabilizers P. M. Quan

THE flaking of paint and the perishing of rubber are regrettably well-known examples of the deterioration suffered by most organic materials after prolonged exposure to air and light. Organic polymers, both natural and synthetic, are subject to this ageing process. Indeed the paint and rubber might be made either from naturally-occurring materials (linseed oil and rubber, respectively) or from synthetics such as poly(vinyl acetate) and styrene–butadiene polymer— the effects observed would be similar. The ageing process involves absorption of atmospheric oxygen. Although different polymers age in different ways, their physical properties will depart further from the initial optimum as more and more oxygen is absorbed. This may result in changed appearance, as when an article becomes discoloured or a transparent material becomes opaque, or, more seriously, in softening, cracking, or crumbling.

To counteract these processes, stabilizers are used. The term 'stabilizer' indicates an additive that prevents or retards deterioration of the substrate. Usually the deterioration is oxidative, so most stabilizers are antioxidants, and more specific terms such as 'antiozonant', 'ultraviolet stabilizer' and 'metal deactivator' indicate effectiveness against the different but interwoven mechanisms by which oxidation can occur.

Stabilizers are used in petrol, foodstuffs, cosmetics, and many commodities other than synthetic polymers. The largest single use is that of antioxidants in rubber. By their nature stabilizers are not sold to the ultimate customers, the public, but are used by manufacturers who convert raw materials (including polymers) into useful articles. The manufacturers usually possess facilities for comparing the effectiveness of one stabilizer against another. If the stabilizer is highly effective, it can command a high price per ton. Stabilizers are therefore specifically designed by the fine chemicals industry for high efficacy (as are dyestuffs). Manufacturers operate internationally and have to be internationally competitive. Their stabilizers find markets because they are necessary for the effective performance of modern synthetic polymers. Indeed at least one polymer with an otherwise very useful balance of properties, polypropylene, would not be manufactured at all, were it not for the stability which antioxidants confer. Figure 1.1 shows the rate of oxygen absorption by polyethylene at 140°, and the effect of adding 0·1 per cent of different antioxidants to the polymer. The characteristic shapes of these curves are discussed later.

Ageing of polymers and the role of stabilizers

The ageing of polymers containing, principally, hydrocarbon chains occurs by a sequence of reactions. The sequence has been established partly by

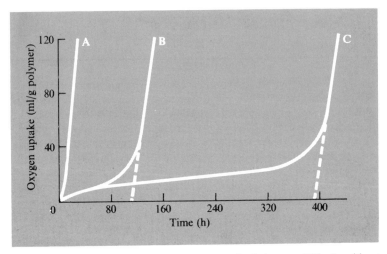

FIG 1.1. Oxygen absorption by high-pressure polyethylene at 140°. A, without antioxidant; B, containing 0·1 per cent of bis(2-methyl-4-hydroxy-5-t-butylphenyl) sulphide, **1.24**; C, containing 0·1 per cent of *NN'*-di-2-naphthyl-*p*-phenylenediamine, **1.19**.

experiments with polymers themselves and partly by analogy with simpler model compounds. The results of modern kinetic studies are consistent with this explanation, which accounts for the anomalies which can arise from the presence of antioxidants in the polymer. A description of these reactions will be useful here for comparing different types of antioxidant.

The first step in the oxidation cycle is the generation of a free radical at some point on the polymer hydrocarbon chain:

$$(\text{Polymer}) \xrightarrow{\text{initiation}} \text{R} \cdot \qquad (1.1)$$

This initiation could be caused by interaction of ultraviolet light with a chromophoric group or impurity in the polymer, or by heating, or by mechanical damage when the polymer is stretched, flexed, or milled. Mechanical damage is quite likely in a viscous polymeric medium since the chains cannot easily move to relieve applied strain. A radical is most likely to be formed at a position where it will be stabilized. This stabilization can be by 'hyperconjugation', at a point of chain branching, or by delocalization, at a position allylic to a double bond (i.e. $C-C=$). Once generated, the radical could undergo further reactions resulting in breaking of the polymer chain or in crosslinking, or it could combine with another radical to form inactive products.

Alkyl radicals are very unstable: the lifetime of radicals generated in polyethylene has been estimated at 10^{-8} s. If oxygen is present, alkyl radicals very readily give more stable alkylperoxy radicals, with lifetimes of about 10^{-2} s. These stabilize further by hydrogen abstraction from the polymer to give hydroperoxides, at the same time regenerating an alkyl radical (eqns 1.2 and 1.3).

$$R\cdot + O_2 \rightarrow ROO\cdot \qquad (1.2)$$
alkylperoxy radical

$$ROO\cdot + RH \rightarrow ROOH + R\cdot \qquad (1.3)$$
hydroperoxide

Compared with their radical precursors, hydroperoxides are stable species (they can in favourable circumstances be isolated as pure compounds). Their concentration in the polymer therefore increases with time. Yet by normal chemical standards these peroxides are very unstable. The danger of concentrating peroxides in distillation residues (e.g. of diethyl ether) is well known. The process (1.4) for manufacturing phenol and acetone in which air is passed into cumene depends on the formation of the hydroperoxide which is then catalytically decomposed.

$$
\begin{array}{ccc}
\underset{\text{cumene}}{\text{Me}\diagdown_{\text{CH}}\diagup\text{Me}} & \xrightarrow{O_2} \underset{\text{cumene hydroperoxide}}{\text{Me}\diagdown_{\text{C--OOH}}\diagup\text{Me}} & \xrightarrow{[H^+]} \text{Me}_2\text{CO} + \text{OH}
\end{array} \qquad (1.4)
$$

Decomposition of the hydroperoxide formed in (1.3) could take place similarly, breaking a polymer chain to give oxygenated products of lower molecular weight which are not free radicals. Possible products from the breakdown of secondary and tertiary hydroperoxides are given in (1.5). It is believed that most of the oxygenated breakdown products found in polymers are formed by such hydroperoxide decompositions.

$$
\begin{array}{ll}
\underset{R'}{\overset{R}{\diagdown}}\text{CH--OOH} & \nearrow \quad \text{RCHO} + \text{R'OH} \\
 & \searrow \quad \underset{R'}{\overset{R}{\diagdown}}\text{CO} + \text{H}_2\text{O}
\end{array} \qquad (1.5)
$$

$$
\underset{R''}{\overset{R}{\underset{\diagup}{\diagdown}}}\text{R'--C--OOH} \rightarrow \underset{R'}{\overset{R}{\diagdown}}\text{CO} + \text{R''OH}
$$

Alternatively the hydroperoxide formed in (1.3) may dissociate to give two free radicals in a reaction which may be monomolecular (1.6), or bimolecular (1.7). In either case two radicals are formed which are capable of abstracting

$$ROOH \rightarrow RO\cdot + \cdot OH \tag{1.6}$$

$$2\,ROOH \rightarrow RO\cdot + ROO\cdot + H_2O \tag{1.7}$$

hydrogen from the polymer to give new macroradicals $R\cdot$.

Taken together, (1.2) and (1.3) represent a self-perpetuating chain-reaction, but one which might not be very harmful because of the rarity of the initiation step (1.1) and because the chain would be eventually terminated by, for example, the combination of two radicals. The breakdown of hydroperoxide is chain-branching or degenerate: it provides a mechanism for increasing the number of oxidation chains. It explains the autocatalytic nature of most polymer oxidations, which was illustrated by the shape of the oxygen absorption curves in Fig. 1.1, and has been confirmed by showing that the rate of oxygen absorption increases sharply if peroxides or hydroperoxides are added to the system. The induction period, or time during which the initial slow build up of peroxides is taking place, is followed by rapid deterioration and is often used as a convenient measure of the useful life of the polymer. In Fig. 1.1, inclusion of antioxidants has increased the induction period at 140° from about 5 hours to 110 hours and 390 hours, respectively.

By interfering with one or more of the reactions listed above and hence terminating the chain reactions, or preventing their initiation, antioxidants can delay oxidation to an extent which might not have been expected from the low concentrations (usually 0·01–1·0 per cent by weight) at which they are used. Antioxidants may be classified in terms of the species with which they are believed to react; these classes are discussed here in order of their practical importance.

Radical-reactive antioxidants

Phenol was recognized to be an antioxidant for natural rubber in 1870, and aniline some forty years later. Aromatic amines and substituted phenols together still form by far the most important antioxidant class. Both types of compound are themselves easily oxidized and it was initially believed that their role was sacrificial, i.e. that they reacted with oxygen before it could react with the polymer. They are indeed consumed as the polymer ages, but this simple explanation could not account for their effectiveness at very low concentrations. It is now realized that direct oxidation is a harmful side-reaction, and that the useful property of both types of compound is their ability (cf. schemes 1.8 and 1.9) to donate hydrogen to free radicals with a facility which may be attributed to resonance stabilization of the radicals to which they themselves give rise. Thus, on hydrogen abstraction;

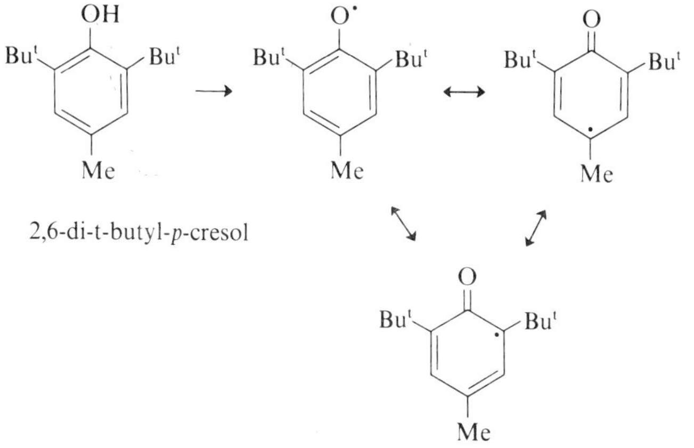

2,6-di-t-butyl-*p*-cresol

(1.8)

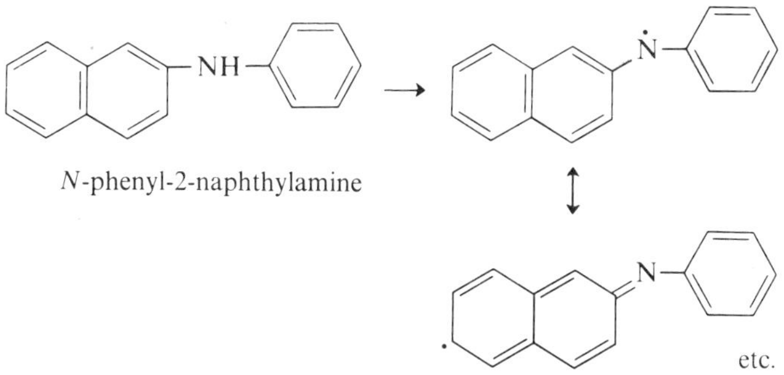

N-phenyl-2-naphthylamine

(1.9)

etc.

Because alkyl radicals react so rapidly with oxygen, their reaction with the antioxidant is probably a rare event, and the antioxidant must react mainly with peroxy radicals. This is shown in (1.10) where the antioxidant is HA, and radicals of the type just depicted are A·.

$$ROO· + HA \rightarrow ROOH + A· \qquad (1.10)$$

Contrast this with (1.3):

$$ROO· + RH \rightarrow ROOH + R· \qquad (1.3)$$

The antioxidant breaks the chain by substituting an aromatic radical A· for the chain-propagating radical R·. Clearly such a process is advantageous only to the extent that A· does not take part in radical-transfer reactions with the polymer. The phenols widely used as antioxidants contain bulky alkyl substituents in one or both of the *ortho*-positions, and their performance in a particular system depends critically on the size of these substituents. To explain this it has been suggested that both the required reaction (1.10) and subsequent harmful radical-transfer reactions of A· will be slowed by steric hindrance, but probably not in exactly the same way. It would not therefore be unreasonable to expect an optimum molecular structure that varied with the polymer substrate.

Further reactions of the antioxidant are complex, frequently leading to highly-condensed and tarry products. To clarify them the controlled oxidation of antioxidants and particularly of phenols has been studied in non-polymeric systems. The highly-hindered phenoxy-radical prepared from 2,4,6-tri-t-butylphenol (**1.1**) is quite stable in the absence of oxygen and can be isolated as a blue crystalline solid. It reacts with oxygen to give the peroxide (**1.2**). Other hindered phenoxy radicals readily dimerize without addition of oxygen. e.g. (**1.3 → 1.4**). They react also with peroxy-radicals to give peroxydienones (**1.3 → 1.5**).

Phenolic antioxidants can therefore deactivate more than one alkylperoxy radical. However, they give rise to peroxides (see (1.10) and compounds **1.2** and **1.5**) which are potential initiators of oxidation. At high temperatures, when peroxides decompose more rapidly, and the difference between the reactivities of R· and A· is less, they become relatively ineffective.

Radicals derived from amine antioxidants can also dimerize (**1.6**) or undergo further oxidation to give, for example, (**1.7**). Nitroxides like (**1.7**) are themselves antioxidants. The further reaction products of aromatic amines seem invariably to be coloured, and therefore to stain the substrate. This defect prevents their use in many applications where their high antioxidant activity would otherwise be welcomed.

       1.6   **1.7**

Peroxide-decomposing antioxidants

An agent which decomposed hydroperoxides to non-radical products would clearly be a useful antioxidant. Preferably it would function catalytically so that it would not itself be consumed. Strong acids catalyse peroxide decomposition, as was mentioned in connection with cumene hydroperoxide (eqn 1.4). They have proved to be very powerful antioxidants in model systems. Unfortunately strong acids are not suitable for inclusion in polymers. Nitrogenous bases can also decompose peroxides, and the high antioxidant activity of the aromatic amines may be partly due to the availability of this additional, secondary, mechanism.

Several commercially important sulphur- and phosphorus-containing antioxidants owe their activity to peroxide-decomposing rather than radical-scavenging mechanisms. These include simple and mixed esters of phosphorus acids with alcohols and phenols, and certain thiols and their metal salts. The most important, however, are long-chain aliphatic esters of β,β'-thiodipropionic acid (**1.8**), particularly the dilauryl and the distearyl esters, known

$$S \begin{cases} CH_2CH_2CO_2R \quad DLTP \quad R = n\text{-}C_{12}H_{25} \\ \\ CH_2CH_2CO_2R \quad DSTP \quad R = n\text{-}C_{18}H_{37} \end{cases}$$

1.8

as DLTP and DSTP. The long carbon chains in these compounds are of physical rather than chemical significance; they assist in making the molecules compatible with polyalkenes and, by adding to the molecular weight of the agent, they minimize its loss by evaporation from the polymer.

Addition of DLTP or DSTP to an oxidizing substrate can produce a slight pro-oxidant effect during the initial stages of oxidation. An explanation recently advanced is that DLTP and DSTP undergo a slow oxidation in the oxidizing substrate to sulphoxides, sulphinic acids, and finally to sulphur dioxide. These oxidations produce radicals which contribute to the oxidation of the substrate. Sulphur dioxide is, however, a very powerful peroxide-decomposer, and is held to account for the net antioxidant effect.

Probably because of these pro-oxidant stages DLTP and DSTP are themselves only weak antioxidants. Their use in combination with phenolic antioxidants is of great interest, however, because effects are obtainable which are much greater than the sum of the effects of the separate components (see Fig. 1.2). This phenomenon, known as *synergism*, can sometimes be observed between antioxidants of the same mechanistic type, but to a lesser extent. Qualitatively, one can understand that the phenol protects against the radicals produced during oxidation of the DLTP, or alternatively, that the DLTP extends the life of the phenol by removing from the system the peroxides which would otherwise give rise to reactive radicals. Combinations of DLTP or DSTP with high molecular-weight phenolic antioxidants in selected proportions, perhaps with addition of an ultraviolet stabilizer, are much used for protecting polyalkenes in their more demanding applications.

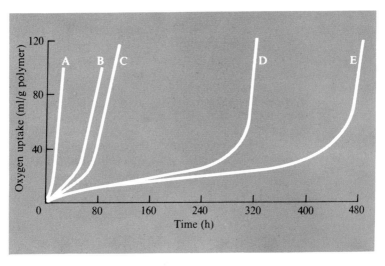

FIG 1.2. Synergism of antioxidants in polyethylene at 140°. A, without antioxidant; B, with 0·1 per cent of DLTP **1.8**; C, with 0·1 per cent of 'Topanol CA', **1.25**; D, with 0·05 per cent of 'Topanol CA' and 0·05 per cent DLTP; E, 0·1 per cent 'Topanol CA' and 0·1 per cent DLTP.

Such combinations are not used in unsaturated rubbers which, except for a few specialized cases, show little response to peroxide-decomposing antioxidants.

Phosphites are rather less efficient than DLTP and DSTP when used in mixtures with phenolics, but they do protect polyalkenes against ultraviolet radiation.

Ultraviolet stabilizers

Sunlight is very important in the degradation of polymers, and the most damaging part of its spectrum is at about $33\,000\ \text{cm}^{-1}$ (300 nm), in the near ultraviolet. At shorter wavelengths most of the light is absorbed by the atmosphere. Exposure to sunlight accelerates normal oxidative ageing and can produce in a few weeks changes which may not be shown for several years by the same polymer kept in the dark.

It is well known that ketones absorb u.v. light in the 300 nm region and can decompose under its influence to give free radicals (1.11).

$$RR'CO \xrightarrow{h\nu} R\cdot + R'\cdot + CO \qquad (1.11)$$

Hydroperoxides are decomposed even more readily.

In a polyalkene undergoing oxidation, ketones and hydroperoxides will be present (eqns 1.3 and 1.5). It is therefore understandable that by converting these species into radicals, the u.v. light will catalyse the oxidation. A pure polyalkene, however, would not absorb u.v. light in the damaging part of the spectrum. There is evidence that oxygen can form chromophoric charge-transfer complexes with the polymer, and these may provide the initial trigger.

Antioxidants previously described retard u.v.-catalysed degradation but because of the severity of the process are not wholly effective. Opaque fillers and pigments used in compounding act as screens reducing degradation and tending to confine it to the surface. Carbon black, much used as a filler because of the excellent physical properties which it imparts to polymers, is very efficient in this respect. Such screens are absent in transparent goods used out of doors, and may otherwise be insufficient. In these cases u.v. stabilizers are added to the polymer. These are transparent to visible light but absorb strongly near 300 nm. Because the successive introduction of conjugated double bonds into a molecule normally produces large increases (10–100 times) in the light-absorption coefficient, it is quite easy to find compounds which will efficiently absorb in the harmful spectral region even when they are present at concentrations of only 0·5 per cent. The difficulty is that such efficient absorbers are just those compounds which are likely to dissociate under the influence of light, or to use the energy absorbed in attacking the polymer, with production of oxidation-initiating radicals. It is necessary that the stabilizer should dissipate harmlessly the energy it absorbs.

This is well illustrated by the benzophenones, which are good absorbers because of the conjugation between the carbonyl group and the two benzene rings. Benzophenone itself (**1.9**) is a powerful catalyst of u.v.-initiated oxidation, whereas 2-hydroxybenzophenone (**1.10**) and its substituted derivatives comprise the most important class of commercial u.v. stabilizers. This difference is ascribed to the keto–enol transformation (**1.10–1.11**) which utilizes the light energy and is subsequently reversed with evolution of heat.

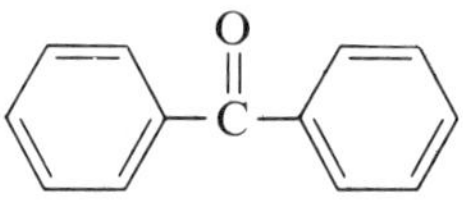

1.9

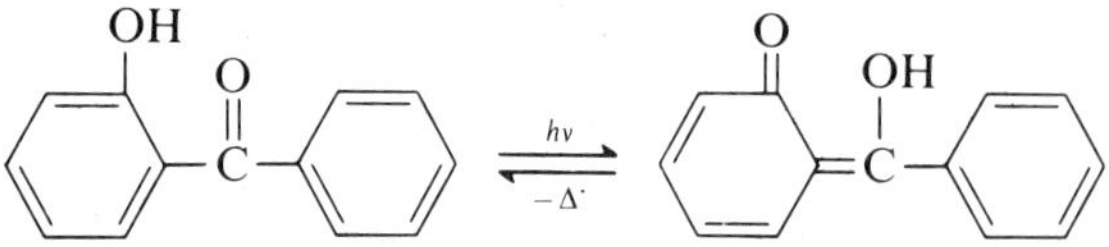

1.10   **1.11**

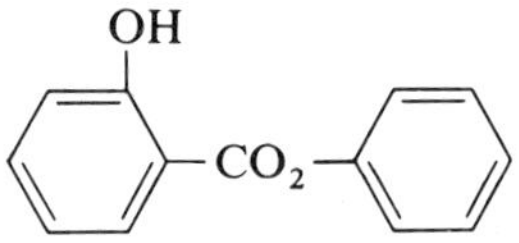

1.12 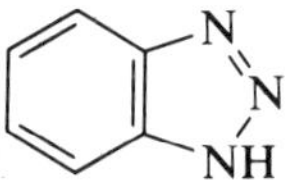   **1.13**

Perhaps fifty different u.v. stabilizers functioning by the absorption mechanism are manufactured throughout the world, and are available under a still larger number of brand names. Most of these can be structurally classified as substituted derivatives of phenyl salicylate (**1.12**), 2-hydroxybenzophenone (**1.10**), or benzotriazole (**1.13**). Some nickel complexes, though only weak absorbers, are also used. Their mode of action is uncertain: they may quench the photoexcited states of molecules responsible for initiating the degradation.

Metal-deactivators

Metal ions, particularly those with more than one available valency state such as copper, iron, and manganese, can seriously accelerate the ageing of polymers. The problem is not well understood, and is made more confusing by the ability of the same metals, in combination with complexing agents deliberately introduced or accidentally present in the polymer, either to become more strongly pro-oxidant or to behave as antioxidants. For example, copper salts are used to stabilize nylon but are particularly deleterious in rubber. Metals may find their way into polymers during manufacture from raw materials, chemical plant, or catalyst residues, and later from compounding ingredients. All these must be carefully controlled.

The normal reagents specifically chelating with metals are of little use as antioxidants in synthetic polymers. Some antioxidants however, in addition to their normal properties, are effective in the presence of metals, particularly of copper. Derivatives of *p*-phenylenediamine and certain of the phenolic antioxidants show this effect, and would be chosen, for example, in cable sheathings over copper wire.

Antiozonants

The atmosphere at ground level normally contains about one or two parts of ozone in a hundred million. Ozone attacks unsaturated rubbers which are under strain by a mechanism quite different from the processes so far described. Attack takes place from the surface of the rubber, and cracks form at right-angles to the applied strain. They can usually be observed on the sidewall of an old tyre, where they are distinguishable by the uniformity of their direction from the random crazing-pattern due to surface oxidation. The mechanism of propagation of the cracks and the relationship of their number and rate of growth to the applied strain is a much-studied physico-chemical problem which has not been solved. Chemically, cracks originate in the addition of ozone to double bonds to form ozonides. It is likely that an ozonized layer will form a surface skin with stretching properties differing from those of the bulk of the rubber. Formation of one crack will locally relieve the strain in the surface of the rubber, making it unlikely that a second crack will form very close to the first. The stress within a crack will be directed to the ends, tending to make it grow, and as it grows a new surface will be exposed to ozone attack.

A defence which works well for statically-stressed rubber is inclusion of a wax with the compounding ingredients. Being only partly soluble in rubber, the wax tends to migrate or 'bloom' to the surface, thus forming a protective skin that is to some extent self-renewing. For dynamic stresses, the technique is harmful since attack is concentrated, and cracks grow rapidly, at points where the skin has broken. For reasons unknown, but probably connected with a tendency to bloom to the surface, the *p*-phenylene diamine class of antioxidants contains many antiozonants active under conditions of dynamic stress. Within this class, compounds of general formula **1.14** are particularly useful.

Me
\
  CH·NH—⟨=⟩—NH—⟨=⟩   (**1.14**; R = alkyl)
/
R

The effects of ozone cracking under dynamic conditions merge into those of the phenomenon known as *flex cracking*. This is an oxidative process accelerated by radicals produced by mechanical damage. Flex cracking responds well to certain amine antioxidants (not necessarily those which are effective against ozone) and less well to some of the phenolic antioxidants.

Fire retarders

Fire retardancy is a very important problem in the polymer industries, not least because the flammability of organic polymers is a barrier to enormous potential outlets. However, fire retarders are not normally classed as stabilizers and mention here will be therefore very brief.

Many proprietary halogen- and/or phosphorus-containing compounds are added to polymers to decrease their tendency to burn after an externally-applied flame has been removed. The effects which these compounds show are greater than would be expected by simple dilution of the flammable with the less-flammable material, and it seems likely that they release radicals which interrupt the combustion reactions taking place in the flame. Examples of synergism exist, particularly between the halogen compounds and antimony trioxide, and mixtures of the two are frequently used. The same principle is exploited in using antimony trioxide as a fire retarder for chlorinated polymers, e.g. poly(vinyl chloride).

To overcome problems of volatility the retardant can be presented as a monomer for incorporation into the network of polyesters, polyurethanes, and epoxy resins. Chlorendic anhydride (perchloronorborn-2-ene-5,6-dicarboxylic acid anhydride) and 'Fyrol 6' are examples of such reactive structures.

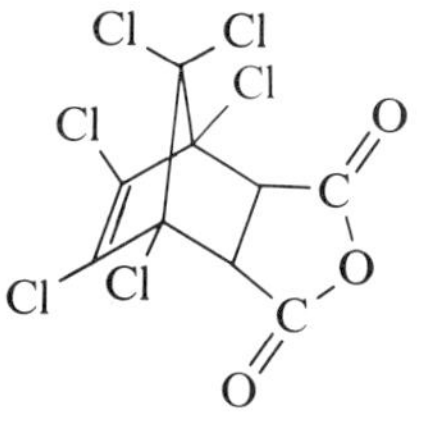

$$(EtO)_2\overset{\overset{\displaystyle O}{\|}}{P}CH_2N(CH_2CH_2OH)_2$$

'Fyrol 6' (Stauffer Chemical Co.)

Chlorendic anhydride

Manufacture of stabilizers

The amines represented by compounds **1.15**, **1.16**, and **1.17** are powerful antioxidants and cheap because of the simplicity of their manufacture.

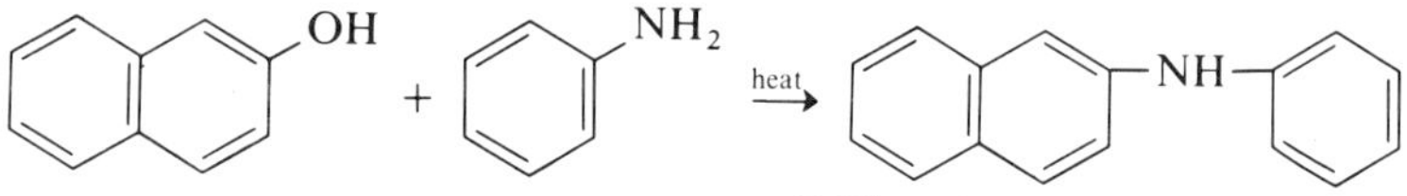

N-Phenyl-2-naphthylamine **1.15**

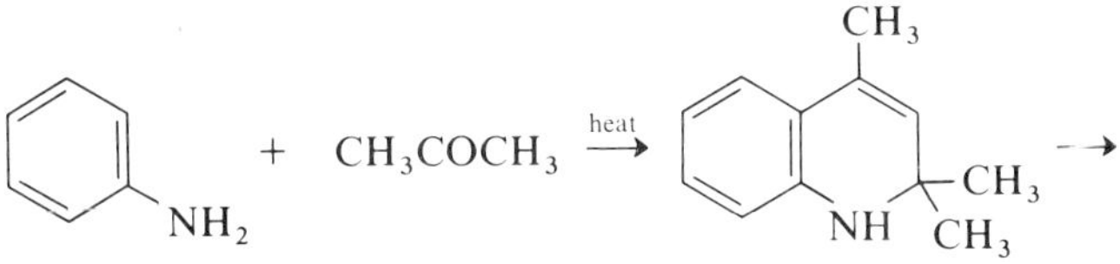

1.16

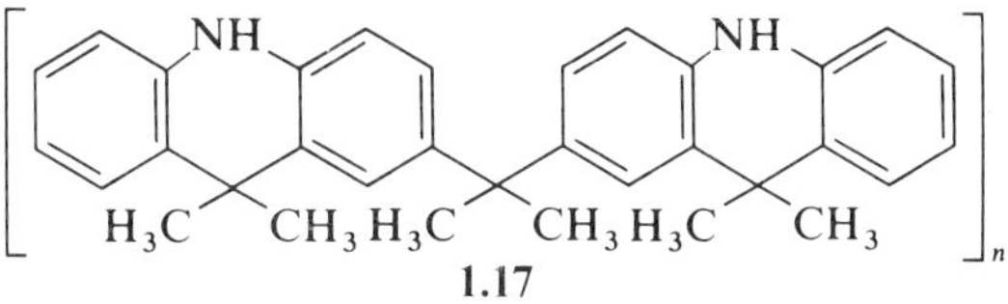

1.17

These products are usually only one reaction removed from chemical intermediates manufactured in high tonnages for a variety of uses. In general the reaction is carried out without solvent. Many of the products are prepared by condensing aromatic amines with aldehydes or ketones (e.g. **1.16**, **1.17**) and comprise mixtures of dimers and trimers with higher polymers. The criterion, of course, is technical efficacy rather than chemical purity, and mixtures may have advantages over pure compounds in this regard, and certainly in cost-effectiveness. Since aromatic amines characteristically stain the substrate and are used only in dark-coloured stocks, slight colour in the products themselves is no disadvantage. It is important however that, if a liquid, the condensate should show no tendency to crystallize on storing at low temperatures, and if a solid, that it should be neither dusty nor sticky, but free-flowing in hoppers and metering devices. It should also disperse readily in the polymer. In the industry, much attention is therefore given to modifying the physical form and particle size of solid products.

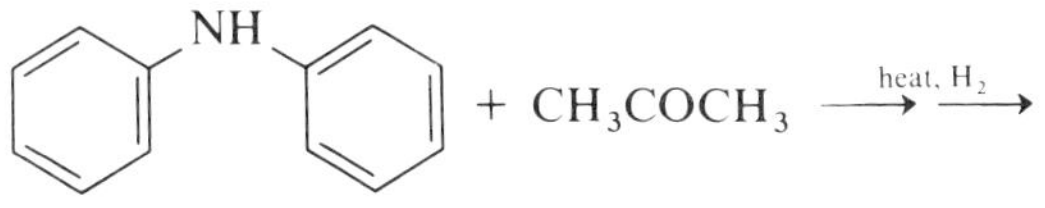

These amines set a cost-effectiveness standard for staining antioxidants. Syntheses requiring several stages or more expensive starting materials can only be justified by specific advantages. Thus the N-alkyl-4-aminodiphenylamines (**1.18**) are used because they are antiozonants as well as powerful antioxidants and NN'-diaryl-p-phenylene diamines (**1.19**) because they are metal inhibitors.

In the presence of acidic catalysts, phenols may be alkylated in vacant ortho- or para-positions with alkenes or alcohols. A wide range of cheap alkylating agents is therefore available. Introduction of the bulky t-butyl group in one or both *ortho*-positions using isobutene (2-methylpropene) is easy, and important in many antioxidant syntheses. It is a reversible reaction,

4-(isopropylamino)diphenylamine
1.18

NN'-di-2-naphthyl-*p*-phenylenediamine   **1.19**

so there are possibilities for recycling side products having the wrong substitution pattern. Some of the alkylated monohydric phenols, e.g. **1.20** and **1.21**, are as inexpensive as the cheap amines, and are widely used in non-staining applications. They are, however, rather volatile; this disadvantage may be overcome by joining together two molecules of a phenol with a vacant *ortho-* or *para*-position. A ketone may be used to form the link, as with **1.22**, but more frequently an aldehyde is chosen so that **1.23** is typical of many compounds on the market. Sulphur dichloride is also used to introduce a linking group (**1.24**). Certain more highly-condensed phenols, **1.25**, **1.26**, are manufactured mainly because of their high activity at low concentrations in polyalkenes.

Aryl phosphites **1.27** are prepared by reacting phenols with phosphorus trichloride.

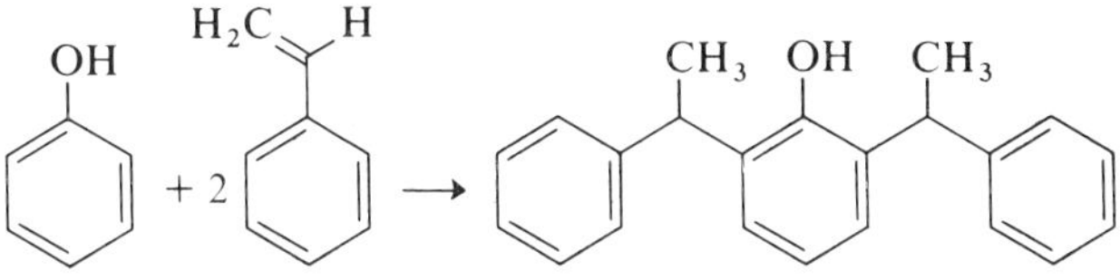

1.20

(and related products)

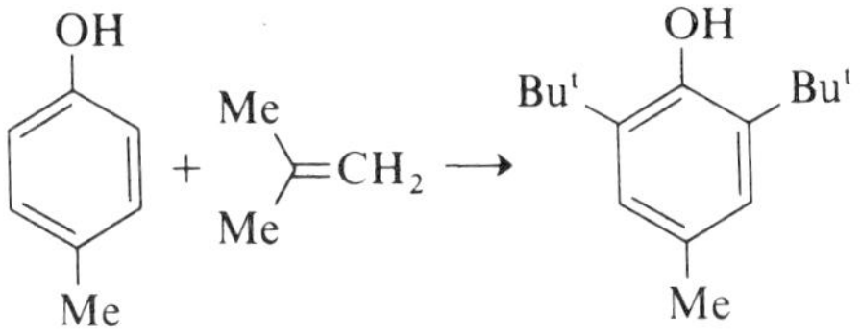

BHT (butylated hydroxytoluene) or
2,6-di-t-butyl-*p*-cresol   **1.21**

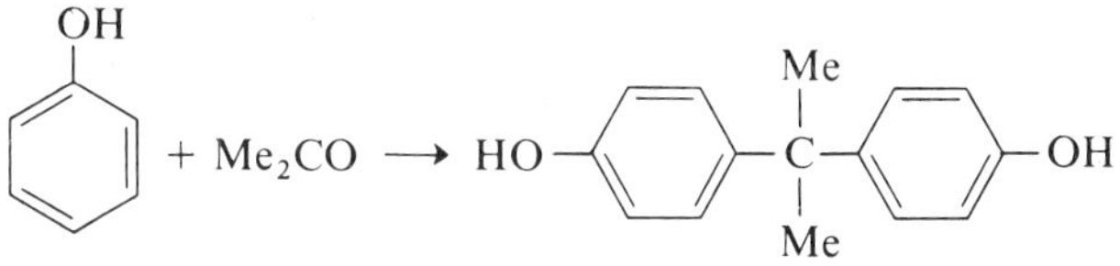

'Bisphenol A'   **1.22**

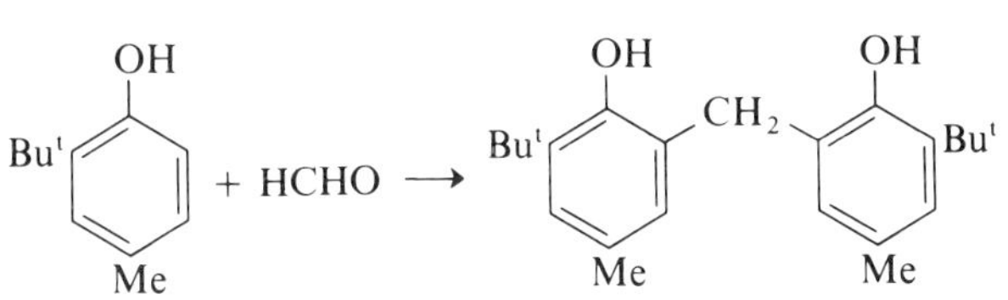

bis(2-hydroxy-3-t-butyl-5-methylphenyl)methane
1.23

bis(2-methyl-4-hydroxy-5-t-butylphenyl)sulphide
1.24

'Topanol CA' (I.C.I.)   **1.25**

'Irganox 1010' (Geigy)
1.26

tris(p-nonylphenyl) phosphite   **1.27**

Thiodipropionic acid **1.28** is prepared by reacting acrylic acid or one of its derivatives with hydrogen sulphide and may be esterified with long-chain alcohols to give DLTP or DSTP. **1.8**.

1.28

The simplest approach to the benzophenones used as u.v. stabilizers is to condense a suitably-substituted benzoic acid, or benzoyl chloride with a substituted phenol using an acidic catalyst, but many other methods are claimed in the patent literature. In general, several stages are necessary and the products require careful purification, and are at the higher end of the antioxidant price range.

The total sales-value for each of the main antioxidant classes used in rubbers is shown in Table 1.1; the price, and prices of probable main starting-materials for one antioxidant of each class are given in Table 1.2. Figures refer to the U.S. market because these are more readily obtainable. They vary with time, but provide a comparative guide.

TABLE 1.1

Sales of rubber antioxidants in the U.S.A. in 1970[†]

Antioxidant class	Examples	Sales ($ million)
Phenylnaphthylamines	**1.15**	5
Aldehyde– and ketone–amine condensates	**1.16, 1.17**	4
Antiozonant p-phenylene diamines	**1.14, 1.18**	40
Non-antiozonant p-phenylene diamines	**1.19**	9
Alkylated or styrenated phenols	**1.20, 1.21**	21
Polyphenols	**1.22, 1.23, 1.24**	28
Miscellaneous	—	4

[†] R. M. Hull and L. G. Parkinson, *American Chemical Society* (*Rubber Chemistry Division*) *Meeting at Chicago*, October 1971.

TABLE 1.2

Prices of antioxidants and of probable starting materials in the U.S.A.

Antioxidant	Starting materials and their prices (US $ per lb)		Antioxidant price ($ per lb)
Phenyl-2-naphthylamine **1.15**	2-Naphthol,	0·35	0·60
	Aniline	0·15	
Acetone-diphenylamine condensate **1.17**	Acetone,	0·06	0·50
	Diphenylamine,	0·26	
N-isopropyl-*N'*-phenyl-*p*- phenylene diamine **1.18**	*p*-Nitrochlorobenzene,	0·29	1·20
	Aniline,	0·15	
	Acetone,	0·06	
NN'-diphenyl-*p*-phenylene diamine	Quinol,	0·82	1·16
	Aniline,	0·15	
Styrenated phenol **1.20**	Phenol,	0·08	0·59
	Styrene,	0·08	
Bis(2-methyl-4-hydroxy-5- t-butylphenyl) sulphide **1.24**	2-t-Butyl-5-methylphenol,	0·54	1·21
	Sulphur dichloride,	0·05	

Choice of stabilizers

Upwards of a hundred structurally-different antioxidants (in addition to u.v. stabilizers) are manufactured throughout the world for use in the polymer industries. There is, of course, some duplication of effects, because new compounds may be introduced when suitable raw materials have become economically available, or to avoid patent infringement, or for other reasons which are not primarily concerned with improved performance. However, a wide choice is desirable because there is as yet no universal stabilizer. Inversion in the relative activities of antioxidants on changing from one polymer to another is frequently found. Within a given polymer the performance of antioxidants will vary with the fillers and other compounding ingredients, and with the conditions of polymer use. Though our understanding grows, many of these effects are correlated merely empirically. In addition, an antioxidant is likely to possess one or more features unrelated to its antioxidant performance which will be critical to its choice for certain applications. Examples of these are:

it may stain the polymer or adjacent materials;

it may adversely affect the processing of the polymer;

it may bloom to the surface with disfigurement and loss of protection;

it may be volatile, and evaporate during processing or at elevated temperatures during use;

it may or may not be accepted in a particular country for use in food containers or wrappings;
because it is a liquid (or a solid) it may be unsuitable to the storage or metering arrangements of certain customers.

Finally (or firstly) the customer will consider the price of the antioxidant: this could be lower than £0·20 per lb. (£450 per ton) or higher than £2·50 per lb. (£5500 per ton). The cheapest is likely to be effective at about 1·5 per cent incorporation, i.e. to cost about £7 per ton of polymer. To be competitive, the most expensive must therefore be effective at about 0·1 per cent incorporation. In these circumstances, the user will select the antioxidant giving the most satisfactory overall result, in the application envisaged.

Polyethylene, the simplest of the high-tonnage plastics, is fairly stable to oxidation. It is commonly protected by cheap phenolic antioxidants, mainly to avoid the degradation that occurs at processing temperatures. Advantage can be shown for use of smaller quantities (0·05–0·1 per cent) of the high molecular-weight phenols (e.g. **1.25**, **1.26**) in conjunction with DLTP, **1.8**. These more-expensive systems are more commonly used in polypropylene which is less stable because of its branched carbon skeleton, and which, because of its higher melting point, is used at higher temperatures than is polyethylene. Polystyrene itself is fairly stable, but phenolic antioxidants are employed, with DSTP preferred to DLTP as a synergist. Acrylonitrile–butadiene–styrene (ABS) resins are major users of antioxidants, principally of BHT **1.21**. All the polyalkenes need u.v. stabilizers, generally of the benzophenone type, with benzotriazoles preferred for polystyrene.

With few exceptions, the organic rubbers contain unsaturated linkages and are very prone to oxidative attack. Resistance to oxidation is also affected by naturally-occurring impurities and by the wide range of chemicals used in rubber processing and vulcanization, many of which show some antioxidant or pro-oxidant behaviour. These complex systems do not show the marked induction periods found with the polyalkenes, nor do they respond to the same low concentrations of antioxidant. Wherever possible, amine antioxidants are used, and at concentrations of 0·5–2 per cent. The antioxidant is matched to the use, so that a tyre would typically contain a heat-ageing antioxidant (**1.15**, **1.16**, or **1.17**) in the casing and an antiozonant **1.18** in the tread and sidewalls. Phenolic antioxidants are used in non-staining applications. The high molecular-weight types do not in general offer any advantage to offset their cost; therefore those in the intermediate price range, e.g. **1.23** or **1.24**, are used. Rubbers are usually compounded with fillers, and the use of u.v. stabilizers is rare.

Poly(vinyl chloride) (PVC) requires specific mention because of its importance and because its degradation is only in part oxidative. Degradation mainly proceeds with loss of hydrogen chloride and development of colour in the polymer. The structure of PVC **1.29** is such that introduction of a double bond in this way makes the next chlorine allylic, and therefore labile. Discrete and growing series of conjugated double bonds are therefore produced, and as they grow their absorption shifts from the u.v. into the visible region. The colour which is observed is due to the chains of seven or more conjugated double bonds.

$$\underset{\textbf{1.29}}{-\overset{\overset{\displaystyle Cl}{|}}{C}HCH_2\overset{\overset{\displaystyle Cl}{|}}{C}HCH_2\overset{\overset{\displaystyle Cl}{|}}{C}HCH_2-} \;\longrightarrow\; -\overset{\overset{\displaystyle Cl}{|}}{C}HCH_2\overset{\overset{\displaystyle Cl}{|}}{C}HCH=CHCH_2- \;+\; HCl$$

The problem is most severe at the high temperatures necessary for processing the polymer. The most widely-used stabilizers are barium, cadmium, and lead salts of carboxylic acids, but dicarboxylates and dimercaptides of dioctyl- or dibutyl-tin are active at lower concentrations. Although they are more expensive they are increasingly used. All these compounds, which are described as 'heat stabilizers', are weakly basic and act as acceptors of hydrogen chloride. They are believed to check the degradation by substituting labile chlorine atoms with carboxylate or sulphide groups. Oxidation of the plasticizers with which PVC is usually compounded catalyses the dehydrochlorination. Conventional antioxidants and particularly bisphenol A (**1.22**) are therefore used in addition to heat stabilizers.

Further details and descriptions of stabilizer systems for other polymers may be found in the short Bibliography.

The future of the industry

New problems arise as the existing polymers find new uses. For example, polyester tyre cord was found to be weakened by agents produced in the rubber surrounding the tyre cord during vulcanization of the tyre : this has led to the introduction of a new product which protects the cord. The industry is also concerned to improve the effects which it presently sells. Antioxidants are reactive molecules freely migrating in an otherwise inert substrate. This can cause problems of staining in adjacent materials, and even of loss of antioxidant by volatilization or during washing. These problems might be overcome by antioxidants which chemically attach themselves to the polymer chains; such work is in progress. Even more fundamental is the constant search for novel structures with stabilizing effects. The mode of action of present products is less than elegant in that they give a sometimes delicate balance of advantage. The time seems ripe for discovery of compounds providing more solidly-based advantages.

Finally, a problem which, unlike most of those posed to the industry, has been a matter of public debate : should disposable plastic cups, bottles, containers and wrapping materials, which can create unsightly litter, contain pro-degradants to accelerate their decomposition? Clearly the decomposition must not be premature, but patents now claim suitable techniques. The cost must be balanced against the advantages claimed. The position could be affected by legislation, and there is evidence of pressure for this in several parts of the world.

2. Pigments and their use in polymers

A. H. BERRIE

PIGMENTS are used in polymers primarily to confer pleasing colour on materials and in this use they compete with suitable dyestuffs. They differ from dyestuffs in not being chemically bound to the material; they lack the sulphonic acid and amine groups present in dyestuffs which effect this binding. Enhancing the aesthetic appeal of the product is important, for example, in decorative paints, many plastics applications, and in synthetic fibres. In all these outlets mass-pigmentation to some extent replaces the more traditional dyeing processes. Printing inks are pigmented, either black or in single colours to provide contrast, to confer legibility, or to produce multi-coloured systems for full-colour printing. Pigments also have other effects: carbon black, for example, is used as a reinforcement for rubber; pigments are used in camouflage; and for identification, for example, in multi-cored cables to improve safety.

A large market thus exists for pigments; sales were of the order of 100 million pounds (50 000 tons) in the USA in 1971 and have been increasing at 12–14 per cent per year in recent years. Hence the manufacture of pigments forms a significant activity in the chemical industry, but the technology involving the use of pigments is much larger, spreading through the paint, printing ink, plastics, rubber, and synthetic-fibre industries.

This chapter discusses what pigments are, how they are made, how and why they are used, and what properties are needed to ensure that they meet the technological requirements of the industries using them, with emphasis on their use in synthetic polymers. It therefore covers an interesting range of structural and physical chemistry and ranges into use–property correlation and technology.

The nature of pigments used in synthetic polymers

The compounds in use today for pigmenting synthetic polymers have been selected from a much greater number originally designed for application in paints and inks. The most important criterion for this selection was the solubility of the pigment in the polymer. Very low solubility or complete 'insolubility' is desirable for pigments if good heat stability and freedom from migration are to be achieved. The pigments in use are usually divided into: (a) inorganic pigments, and (b) organic pigments.

Table 2.1 compares their properties in synthetic polymers.

Unlike organic pigments, many inorganics are prepared by simple aqueous double-decomposition processes. As a result, undesirable impurities may be present. Soluble heavy-metal salts may cause polymer degradation during

TABLE 2.1

Comparison of properties of pigments

	Inorganics	Organics
Shade range	Wide, occasionally dull	Very wide, can be very bright
Covering power	Weak to moderate	Moderate to very strong
Transparency	Opaque	Transparent to moderately opaque
Cost	Generally cheap—but this may be offset by relative weakness compared with organics	Moderate to very expensive
Heat stability	Good, with few exceptions	Vary from poor to good
Migration properties	Non-migratory, do not 'bleed' or 'bloom'	Vary: some non-migratory, some 'bleed' and 'bloom'
Light fastness	Good	Moderate to good
Dispersibility	Harder, tend to be difficult to disperse	Generally softer textured, easier to disperse
Toxicity	May contain toxic heavy metals	Generally non-toxic; need evaluation

processing; and the presence of ionic compounds in general may reduce the electrical resistance of the pigmented polymer, making it unsuitable for use as an insulator.

The above comparison shows that in many ways inorganic and organic pigments are complementary. Thus, if maximum light-fastness is required, only inorganics (or a few carefully-selected organics) can be used. If, however, the object is to produce a strong, bright shade irrespective of light-fastness properties, an organic pigment should be used. Occasionally, as a compromise in properties, a mixture of inorganics and organics is used.

Overall, the inorganics, dominated by titanium dioxides (white) and carbon blacks, far outweigh the organics. Even for coloured pigments only, inorganic usage is greater than organic. However, with research continually providing improvements, the use of organic pigments in synthetic polymers is expected to increase more rapidly in the future.

Inorganic pigments

Titanium dioxide pigments. These are by far the most important and widely-applied pigments, accounting for over 50% of total pigment usage in synthetic polymers. They are manufactured either by the controlled hydrolysis of titanium (IV) sulphate or by burning titanium tetrachloride in air. The rather sophisticated 'chloride' process has reached full production scale only recently and after considerable processing troubles had been overcome.

Titanium dioxide ('titania') pigments fall into two distinct classes, rutile types and anatase types, which differ in crystal structure. Anatase pigments are photochemically active: they accelerate the photodegradation of polymer in which they are incorporated and of any coloured pigments present. This is undesirable and may be partly overcome by surface-treating the pigment particles with small quantities (~ 1 per cent) of other inorganic oxides, such as alumina. Rutile is much less photoactive; its activity may be completely eliminated by treatment with alumina. Indeed, treated rutile pigments may actually confer photochemical stability on polymer media. Rutile is the preferred form of titania because of its higher opacity and photochemical stability. Anatase grades are used, however, when the object is to obtain a particularly pure white tone.

Carbon black pigments. These are widely applied and account for about 25 per cent of the total weight of pigments used in synthetic polymers. They are manufactured by burning natural gas or oil in insufficient air for total combustion. Carbon blacks have excellent technical properties: they are heatfast, non-migrating, strong, and cheap. They are not only very durable in themselves: they also confer durability on the medium in which they are dispersed. With carbon blacks, particle size is very important. Their visual effect depends on light absorption; scattering, if it is allowed to occur, returns some light to the observer, and lightens the shade. To avoid scattering, particle size must be well below the wavelength of the incident light. A typical carbon black from natural gas would have most particles in the range 5–50 nm.

Coloured inorganic pigments comprise a wide range of structures. The most important are discussed here.

Cadmium sulphides and *sulpho-selenides* are prepared by addition of an alkali sulphide or a mixture of sulphide and selenide to an aqueous solution of cadmium sulphate. The crystal or particle size governs the exact shade and brightness produced and precipitation conditions are therefore controlled. Cadmium sulphide can vary in shade from lemon yellow to orange and the sulpho-selenides provide shades ranging from orange through red to maroon. The deeper shades are obtained by increasing the proportion of selenium. The cadmium pigments combine adequate strength with good opacity while being fast to light and non-migratory; their heat resistance is outstanding. Their main disadvantages are their high cost and dullness in 'reduced' shade (i.e. diluted with titania) and these greatly limit their application.

Lead chromates are prepared by simple aqueous double-decomposition reactions from, e.g. lead nitrate and sodium chromate. The shade depends on the crystal structure,

i.e. orthorhombic   →   greenish yellows

monoclinic   →   yellows, oranges

tetragonal   →   oranges, scarlets

The crystal form obtained is controlled by the precipitation conditions and by the addition of other ions such as sulphate and molybdate. Lead chromate pigments are cheap and have good opacity and adequate strength. The major defect of the lead chromate pigments is, of course, their toxicity which prevents their application in toys and food wrappings. They have other technical limitations, however, including poor heat- and light-fastness: they tend to darken above 240°C in organic media or on prolonged exposure to daylight. They are, moreover, readily attacked by atmospheric sulphur dioxide. Some defects, in particular the poor heat fastness, have been partly overcome by the introduction of grades in which the pigment particles are suitably coated.

Iron oxides may be purified ground ores, or synthetic products. They provide a wide range of shades—yellows, oranges, reds, browns, purples, and blacks—depending on the state of oxidation, degree of hydration, and crystal structure. Iron oxide pigments are cheap and non-toxic if sufficiently pure. They have good opacity while some of them, in particular the synthetic products, are relatively strong. The calcined oxides have good heat-fastness, the hydrated types less so; both types are non-migratory while being out-standingly durable. Their main limitation is their general dullness of shade and their moderate strength.

Coloured organic pigments

Virtually all organics used to pigment synthetic polymers are synthetic. They are generally manufactured from a very few aromatic compounds, notably hydrocarbons, usually referred to as *primaries*. These are obtainable from coal tar or, more recently, from the petrochemicals industry. These primaries are first converted by standard, classical, organic chemical reactions, into a wide variety of *intermediates*. Further reaction of intermediates taken singly or in combination then yields the *pigments*.

The most important organic pigments used in synthetic polymers are as follows.

Phthalocyanines constitute the most important and widely used group of organic pigments. Their discovery and development has received wide attention (p. 85). They are the technical leaders in the shades ranging from blue to green, because their properties are outstandingly good: they are non-toxic, very strong and relatively cheap. With the exception of α-copper phthalocyanine (see below), they have excellent heat- and light-fastness and do not migrate.

The most important of these pigments is copper phthalocyanine (**2.2**). It is prepared by heating together, in a solvent, phthalic anhydride (**2.1**), urea, copper(II) chloride, and a catalyst.

Copper phthalocyanine is obtained in two different crystalline modifica-tions, α-form (reddish-blue) and β-form (greenish-blue), depending on the

2.1

self-condensation
⟶
of four units

2.2

conditioning of the crude product obtained directly from the reaction. Of the two, the α-form is thermodynamically the less stable, and in the presence of organic solvents is converted into the β-form, generally in a coarsely-crystalline non-pigmentary state, with consequent shade-change and loss of colour intensity. This lack of stability can give rise to poor colour-fastness during hot processing, where the polymer or plasticizer may act as a solvent. It has, however, been found possible to stabilize the α-form by introducing a small amount of chlorine (< 1 atom per molecule) into the pigment molecule; this may be done by replacing some of the initial phthalic anhydride by the monochloro-derivative.

Other copper phthalocyanine pigments of importance are the polyhalogeno-derivatives, prepared by the direct chlorination and/or bromination of the parent compound. These are bright greens with excellent application and fastness properties.

Azos are widely used and provide strong, bright, yellow to red, shades, often at a moderate price, while their application and fastness properties vary from moderate to good. They are large complex molecules, reflecting the need for very low pigment solubility in organic media. Particular types of azo-pigment commonly used in synthetic polymers are discussed below.

Azo toners are the insoluble salts of azo compounds containing carboxylic or sulphonic acid groups; e.g. (**2.3**, yellow) and (**2.4**, red). Although very insoluble in organic media, these compounds are often sensitive to wet conditions; their light-fastness is generally only moderate.

2.3

2.4

Coupled bisazos, e.g. (**2.5**, orange) are generally prepared from a tetrazotized benzidine reacted with a pyrazolone or acetoacetarylamide coupling-component. The pigments often have barely adequate insolubility while light-fastness varies from poor to moderately good.

2.5, R = phenyl

Condensed bisazos, e.g. (**2.6**, red) have technical properties superior to those of the previous class (e.g. **2.5**). They are prepared by condensing a mono-azo compound containing a carboxylic acid group with a diamine.

2.6

In some *mono-azos*, insolubility is achieved by incorporating a multiplicity of substituents, generally amide groups, into the pigment molecule, e.g. (**2.7**, red). Such products often have very good technical properties with the exception of strength which is generally inferior to that of other classes of azo pigment.

2.7

Vats and other polycyclic aromatic pigments are, in fact, selected vat dyes, or closely related compounds, specially prepared and conditioned to meet the requirements of pigments as regards chemical purity, crystal modification, and particle size. These series of pigments cover a wide shade range, as illustrated in examples **2.8–2.12**.

(**2.9**, perinone orange)

(**2.8**, flavanthrone yellow pigment)

(**2.10**, perylene red)

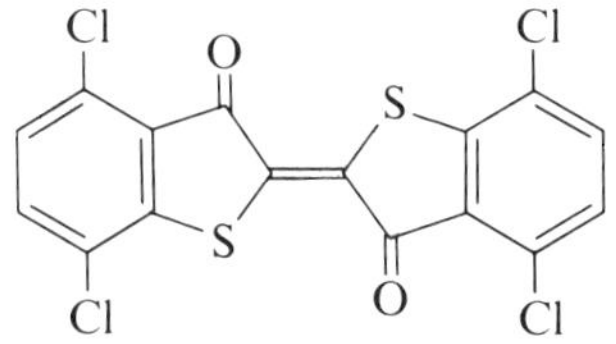

(**2.11**, thioindigoid bordeaux)

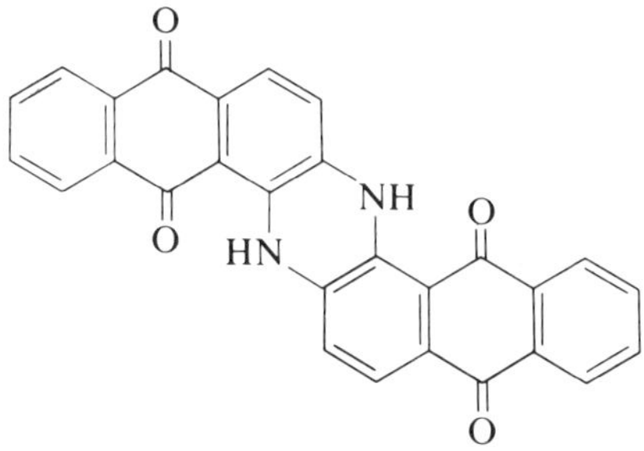

(**2.12**, indanthrone, blue pigment)

Further (non-vat) coloured polycyclic compounds which are applied as pigments in synthetic polymers are the dioxazine (**2.13**, violet), quinacridone (**2.14**, red to violet, depending on physical form) and the isoindolinones (**2.15**, yellow for R = 1,4-phenylene, orange for R = *p,p'*-diphenylene, and bluish-red for *m,m'*-dimethoxy-*p,p'*-diphenylene).

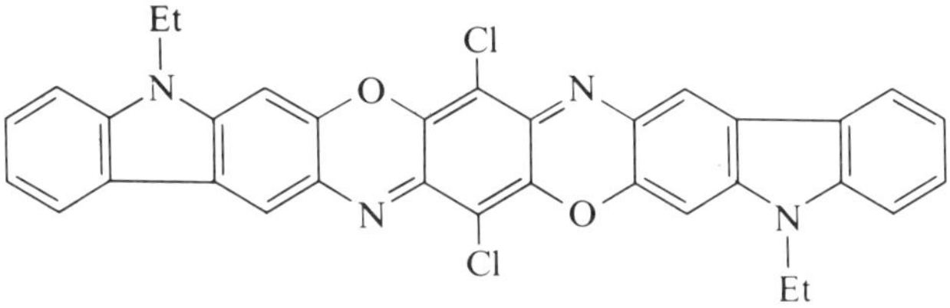

2.13

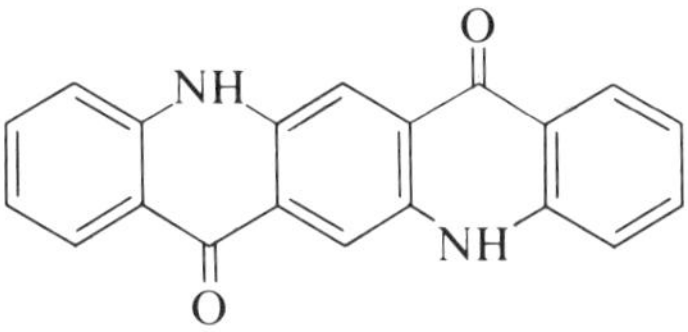

2.14 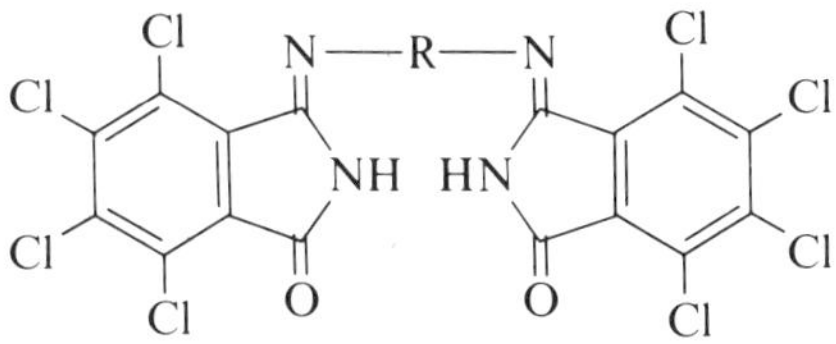   2.15

All these pigments, in particular the quinacridones **2.14** have excellent technical properties, but they are manufactured by multistage processes and are therefore very expensive.

The function of pigments in synthetic polymers

Pigments fulfil a number of functions apart from the main one of conferring colour. Thus, carbon black is used as a stabilizer in many plastics applications where extreme durability is essential. In part the carbon absorbs and screens out harmful radiation and in part it acts as an antioxidant by trapping free radicals and breaking oxidation chains. Pigments may also be used to alter mechanical properties; fillers do this by influencing polymer crystallization. The best known is the reinforcing effect of carbon black in rubbers for tyre manufacture.

The major function of a pigment in a synthetic polymer is, however, to alter the appearance, i.e. the optical properties. The appearance of a pigmented material depends on a combination of light reflected from the surface and light which has penetrated the pigmented material, undergone various modifying processes and then re-emerged.

Surface reflection

When light strikes the surface of a solid, part is reflected while the remainder penetrates the surface and undergoes refraction. The angles of the reflected and refracted beams, f and r respectively, are determined by the angle, i, of the incident beam according to the law of mirror reflection (eqn 2.1) and Snell's law (eqn 2.2), respectively (Fig. 2.1).

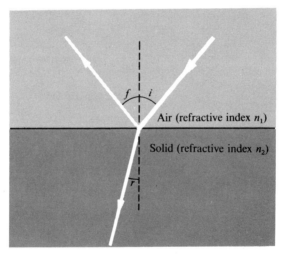

FIG 2.1.

$$f = i \qquad (2.1)$$

$$\sin i/\sin r = n_2/n_1 \qquad (2.2)$$

The fraction of light reflected (ρ) varies with i, n_2/n_1, and the plane of polarization of the electric vector of incident light according to the Fresnel equations. Since light is generally unpolarized, the last factor may be ignored, in which case ρ may be said to increase both with i and with n_2/n_1.

The smoothness of a surface has a profound effect on its appearance, Fig. 2.2.

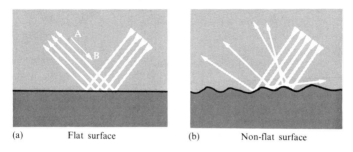

(a)   Flat surface   (b)   Non-flat surface

FIG. 2.2.

In Fig. 2.2a all incident light is reflected in the same direction (mirror or specular reflection). Surface luminance is a maximum viewed in the direction A–B, and zero in all other directions, i.e. the surface is glossy. In Fig. 2.2b, on the other hand, the incident light is reflected in different directions (diffuse reflection). If the surface diffuses light perfectly, surface luminance is the same irrespective of the direction of viewing, i.e. the surface is matt. In a pigmented synthetic polymer the smoothness and hence the degree of gloss is controlled mainly by the smoothness of the moulding or calendering surface used in shaping the polymer. Usually high gloss is desired. A matt effect may be deliberately produced when desired by the incorporation of a high proportion of a filler such as silica in the medium; the product then has microscopic surface irregularities, and behaves as illustrated in Fig. 2.2b.

The degree of gloss affects the colour of a pigmented medium. Thus, in Fig. 2.2a an observer looking along a normal to the surface will see only emergent light, while in Fig. 2.2b he will observe both emergent and reflected light, the latter having approximately the same spectral characteristics as incident light. Thus, in white light, an increase in gloss will lead to a darkening in the appearance of the pigmented medium, while a lightening will result from a decrease in gloss.

Processes occurring within the pigmented material

A fraction $(1 - \rho)$ of the incident beam of light penetrates the pigmented medium. On encountering a pigment particle, the beam may be split into several beams travelling in different directions, a process known as *light scattering*. Further, the light may lose a fraction of its intensity, i.e. *light absorption* may occur.

Light absorption. The visible spectrum extends from about 400 to 700 nm $(25\,000$ to $14\,000\,cm^{-1})$. Different wavelengths within this range appear as different colours. If a compound absorbs no light it appears white or colourless, while if it absorbs throughout the spectrum it will appear grey or black. When, however, a compound absorbs only at certain wavelengths while transmitting the rest, the transmitted light, and hence the compound, will be coloured as indicated in Table 2.2.

TABLE 2.2

Colour of compound and of transmitted light after light absorption

Wavelength absorbed (nm)	Colour of light absorbed	Colour transmitted
400–430	Violet	Greenish yellow
430–490	Blue	Yellow to orange
490–510	Blue-green	Red
510–530	Green	Purple
530–560	Yellow-green	Violet
560–590	Yellow	Blue
590–610	Orange	Greenish blue
610–700	Red	Blue-green to green

The absorption of visible light by a compound involves the promotion of an electron from a ground-state to an excited-state orbital. Absorption of energy is quantized. Therefore this excitation involves absorption at a discrete wavelength. The wavelength and hence the colour is determined by the difference between the energies of the two orbitals.

The electronic transitions most commonly associated with the light absorption in coloured inorganic pigments are of two main types:

(1) Ligand-field (d $\rightarrow$ d) transitions localized on the metal atom; e.g. chromic oxide (Cr_2O_3, d^3 octahedral) is green owing to absorption associated with a $t_{2g} \rightarrow e_g$ transition while the colour of cobalt blue ($CoOAl_2O_3$, d^7 tetrahedral) is associated with an $e_g \rightarrow t_{2g}$ excitation. The energy difference between the donor and acceptor orbitals depends largely on the ligand-field intensity.

(2) Charge-transfer (ligand $\rightarrow$ metal) transitions; e.g. chromates (CrO_4^{2-}, d^0 tetrahedral) whose yellow-to-scarlet colour is associated with a ligand-$\pi \rightarrow$ metal-d_e transition, and cadmium sulphide and selenide (CdS and CdSe, d^{10}) where the yellow-to-red colour is due to absorption associated with a ligand-$\pi \rightarrow$ metal-5s or -6s excitation. In these cases, the energy-difference and hence the colour will depend on the ionization potential of the ligand, the electron affinity of the metal ion and the coulombic attraction between the promoted electron and the 'positive hole' left behind on the ligand.

In the case of organic pigments, the situation is both simpler and more complicated: simpler, since the transitions associated with light absorption are virtually all of the $\pi_1 \rightarrow \pi_1^*$ type, i.e. highest occupied π molecular orbital (MO) to lowest empty π^*MO; more complicated since the factors which determine the energies of the π_1 and π_1^*MOs can be very complex. The most important of these factors are:

(1) Extent of delocalization: increased π-electron delocalization generally diminishes ΔE [i.e. $E(\pi_1^*) - E(\pi_1)$] leading to an increase in the wavelength of absorption;

(2) Steric distortion in the π-electron system: this produces a wavelength shift, the direction of which depends on which part of a given system is distorted;

(3) The presence of substituents, particularly strongly mesomeric ones: these generally produce an increase in the wavelength of absorption, the effect being particularly marked when there are substituents of opposing polarities in the system.

Electronic excitation and colour have been discussed so far in terms of single molecules. However, in solid-pigment particles, molecules are in fact closely aggregated. In such conditions, molecular energy levels may be perturbed. The direction and extent of this perturbation depend on the relative orientation of the molecules in the solid, i.e. on the pigment's crystal structure. The effect of crystal structure on the energy of electronic excitation and hence on colour can be seen in many cases; e.g. lead chromate varies from greenish-yellow to scarlet depending on crystal form; and copper phthalocyanine varies from reddish- to greenish-blue depending on the way in which it crystallizes.

Light scattering. When a beam of light strikes a pigment particle, the processes which occur depend on the size of the particle. There are three cases: particles much larger than the wavelength of incident light; particles much smaller than the wavelength; particles similar in size to the wavelength.

If the particle is much larger than the wavelength of light, then the interaction is governed by the law of mirror reflection (eqn 1), Snell's law (eqn 2) and the Fresnel equations. The extent to which unpolarized light is deflected from its original path, i.e. scattered, will therefore depend on the ratio of the refractive index of the pigment to the refractive index of the medium.

The refractive index of a compound varies with wavelength, particularly in a region where the compound absorbs light, Fig. 2.3. The rapid variation of n in the absorbing region is called 'anomalous dispersion'. The scattering power of a pigment depends on its refractive index; therefore large variations in scattering power are to be expected in the region of an absorption band. In white pigments, which do not absorb in the visible, this effect is unimportant. However, with coloured pigments, anomalous dispersion will lead to certain wavelengths being scattered more strongly than others. The

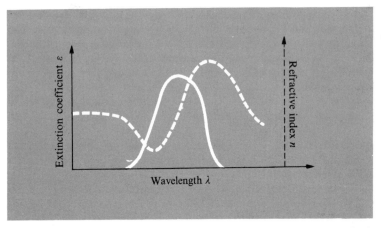

FIG. 2.3. The continuous line is the absorption curve (ε versus λ); the broken line is the dispersion curve (n versus λ).

scattered light thus has a different spectral distribution from the incident light. For example, because of the increase in refractive index near the absorption band more light is reflected in that region and Malachite Green crystals that appear green by transmitted light appear red by reflected light.

Finally, the scattering efficiency of an individual particle within this size range is little affected by particle size. However, for a given weight-concentration of pigment in a medium, a decrease in particle size will increase the number of particles and hence the total number of interactions, leading to more efficient scattering.

For particles much smaller than the wavelength of light, scattering is governed by Rayleigh's law. However, the scattering intensity is very low and for practical purposes may be ignored.

Between the two above extremes there is the intermediate so-called 'diffraction region' in which particle size is of the same order as the wavelength of light. The appropriate theory is complex; however, within this region light scattering is particularly intense and reaches a maximum with particles whose size is approximately half the wavelength of light—about 200–400 nm. As before, scattering intensity is related to other parameters, in particular to the ratio $n_{pigment}/n_{medium}$ and, if there is anomalous dispersion, to λ.

These findings have important application in practice. Thus, to obtain maximum scatter from a given pigment, a particle size should be chosen within the range 200–400 nm. If, on the other hand, low scattering is required, most of the particles should be well below 100 nm in size.

The effects of absorption and scattering. The major optical effect of light absorption by pigments in a synthetic polymer is the generation of colour.

Further, if light absorption is strong and spread over a substantial part of the spectrum, the pigmented synthetic polymer will be opaque.

The major effect of light scattering is to confer opacity on the pigmented material. Thus, a non-scattering pigment in a synthetic polymer will give a transparent effect (unless it absorbs strongly over a wide range of wavelengths) while an increase in scattering power will render the material first translucent and then opaque.

Scattering may also modify the colour produced by light absorption. That this happens is due mainly to incident light being scattered backwards by pigment particles near the surface and re-emerging from the material relatively unmodified by absorption. When incident and scattered light have the same spectral distribution, then in white light increased scatter will lighten the shade. When, however, there is anomalous dispersion, an increase in scattering may lead to a change in hue as well as a lightening in shade.

White pigments function by scattering only; any absorption would lead to darkening or development of colour. Black pigments function entirely by absorption; scattering if it occurred would lead to a lightening of the shade and hence give a less intense black. Coloured pigments function by absorption alone (especially when transparency is required) or by a combination of absorption and scattering.

Re-emergence of light from pigmented material

After undergoing numerous interactions with pigment particles, a fraction of the light which initially penetrated the pigmented material returns to the air–medium interface. Here it undergoes reflection and refraction as before according to the law of mirror reflection (eqn 2.1), Snell's law (eqn 2.2) and the Fresnel equations. In this case, however, a much greater fraction of light is reflected. Thus, for internal, diffuse, non-polarized light in a medium with a refractive index of 1·5 (n_{medium} generally lies within the range 1·3–1·7) only 40 per cent of the light reaching the surface emerges, the remaining 60 per cent being reflected back into the material. This compares with 9 per cent reflection and 91 per cent penetration for external, diffuse, non-polarized light. Finally, the emergent light 'mixes' with incident light reflected directly from the surface. It is this mixture which determines the appearance of the pigmented material.

Methods of applying pigments

Making a pigment and making sure that its properties are desirable are mainly the province of the chemist. A pigment, however, is not used by itself— it is incorporated in a material. The way in which it is so incorporated to best effect is technologically very important. This problem and some of the methods used are briefly discussed in this section.

Pigments are insoluble in the medium in which they are used. They produce colour by existing as discrete particles in the medium, selectively absorbing and reflecting specific light energy. It is important to achieve even coloration of the material. To this end the pigment must be adequately dispersed: particles must be smaller than can be seen by the human eye; and they must be evenly distributed throughout the medium. Powder fineness of the pigment is not a reliable criterion of dispersibility. For example, consider trying to grind coke dust into a medium to simulate the effect of a compressed carbon-black pigment: the two materials will behave very differently. Nevertheless, the major aim in pigment application is to reduce soft aggregates in pigment powders, probably in the 50–100 μm range, until the majority of the particles are a fraction of a μm, and to surround each particle by the medium.

Various methods of dispersion exist, depending broadly on whether the medium is a solid or liquid at room temperature. Usually several stages are involved. In liquid systems such as paint or printing inks, or in plastics where a liquid monomer, plasticizer, or a fluid resin can be used as the pigment carrier, it is common practice first to make a pigment-paste concentrate. The pigment aggregates are broken down by milling. The objective is to pass the pigment aggregates through high shear gradients; this breaks down the aggregates and causes the new pigment surfaces to be wetted with medium. A three-roll mill, a Z-blade mixer, or a ball mill is used to disperse the pigment. This is a batch process and not very efficient. More recently colloid mills, sand mills, or cavitation mixers have come into use; they get away from batch-wise milling and permit more sophisticated in-line continuous-milling operations. In all cases, sufficient pigment is added to the vehicle to give a stiff paste mix, only just fluid enough to be handled on the milling equipment used. Some rough guide to the maximum proportion of pigment to vehicle can be obtained from oil-absorbtion data (i.e. the weight of pigment just wetted by 100 g of linseed oil). The paste mix is passed through the mill a sufficient number of times (or ball milling is prolonged) until it passes a standard test procedure.

The pigment-paste concentrate is then diluted with additional medium to reach the pigment level required in the finished product. The rates of dilution and agitation have to be controlled to avoid risks of flocculation or re-aggregation of the pigment particles. Some media, notably alkyd resin solutions and esters, are better wetting and dispersing vehicles for many pigments than are others, e.g. hydrocarbons. The storage stability of the pigment paste is also important and so, obviously, is its compatibility with the system in which it is to be used.

Where a solid system, such as rubber, plastic, or textile fibre, is being pigmented it is frequently undesirable to introduce a liquid component as a pigment-dispersing vehicle because of the effects of the vehicle on the properties of the final solid. In addition, pigment pastes are not as clean and convenient to weigh and blend with solid polymers as is a powdered pigment, or a pigment master-batch made by dispersing excess of pigment into resin or polymer and grinding. Many solid polymers require incorporation of several additives, of which the pigment is only one. These additives include stabilizers of various types, curing agents, lubricants, and fillers; all are needed to produce a serviceable article at economic cost. It is desirable to incorporate the pigment at the same time as the other solids. Various types of equipment are used to mix and disperse these additives into the polymer while it is softened to a viscous plastic state by heating, as indicated in Fig. 2.4. Route 1 is frequently used for colouring rubber, and for some

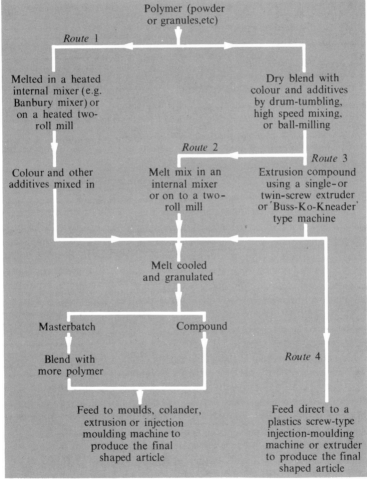

FIG 2.4. The pigmentation of solid polymers.

thermoplastics, route 2 for colouring thermosetting plastics, and some thermoplastics, route 3 for many thermoplastics, and route 4 for thermoplastics and some synthetic fibres.

The pigmentation of certain synthetic fibres, e.g. acrylates and viscose rayon, which are produced by spinning from a solvent solution rather than from the molten polymer, offers special problems. In these cases a concentrated dispersion of pigment has to be prepared in the appropriate solvent or the polymer solution, probably by ball milling, and blended with the bulk of the polymer solution prior to extrusion through spinnerets into the fibre-coagulating bath.

Properties required of a pigment

In selecting a pigment suitable for application in a synthetic polymer, one must take into account:

the synthetic polymer employed;

the method of pigment incorporation;

processing of the pigmented polymer subsequent to incorporation;

the conditions likely to be encountered by the pigmented polymer and the use to which it is put during its lifetime.

The pigment properties most directly relevant to the above considerations are discussed below.

Dispersibility

Good pigment dispersion in the mass pigmentation of synthetic polymers is necessary not only to avoid the presence of streaks or specks of pigment in the product but also to ensure the most efficient use of the pigment's power to absorb and/or scatter light. Poor dispersion may have serious effects: for example, it may lead to electrical failure in a plastic insulator; to mechanical failure in fibres or thin films; or to filter or spinneret blockage in spinning operations. That a good dispersion should be readily achieved is also important, partly to reduce power usage at the mixing stage and partly since the machinery and processes generally used in the industry are relatively inefficient.

Commercially-available pigment powders consist generally of fine particles. The majority cohere loosely in large air flocculates or aggregates. The way these are broken down and the individual particles are surrounded with an envelope of the medium, has already been described in the preparation of pigment pastes. In the polymer the wetted pigment particles have to be distributed evenly throughout the mass of the material. The ease with which these processes occur, i.e. the dispersibility of the pigment, depends on a number of factors including the degree of flocculation and/or aggregation of the individual particles and their surface condition. These features may be improved by adjusting the conditions of physical treatment of the pigment, e.g. how it is precipitated, milled, and dried.

Much attention has been paid in recent years to the preparation of easily-dispersible pigments. In principle the individual pigment particles may be coated with a hydrophobic resin during manufacture; this prevents the formation of difficultly-dispersible aggregates on drying and wets the pigment particles. After mixing and during subsequent hot processing, the resin dissolves or disperses in the synthetic polymer leaving behind the wetted, dispersed pigment particles. Pigments prepared in this way have many technical advantages, e.g. excellent dispersion combined with a low speck level and consistent high strength, readily achieved in simple equipment. On the other hand, the use of a given resin may limit the compatibility of the pigment to a smaller range of synthetic polymers, and the smaller scale of production will lead to higher costs.

Heat stability

This refers to any shade changes occurring due to processes that involve heating the pigment–synthetic-polymer system. Such changes may be caused by:

A change in the physical form of the pigment, for example, flocculation, partial or complete dissolution, crystal growth, or a change in polymorphic form. Such changes may affect the light absorption and light-scattering properties of the pigmented polymer and hence its appearance.

Thermal instability of the pigment leading to partial or complete decomposition. This will lead to a loss in colour intensity or, if the decomposition products are coloured, to a change, generally a dulling, of shade.

Chemical reaction between the pigment and the synthetic polymer itself, additives such as antioxidants or trace impurities such as residual catalysts. Changes in shade may result.

Heat stability thus does not depend on the pigment alone; it is a function of the pigment–medium system as a whole. Further, it will vary with the processing conditions, e.g. the time and temperature of a moulding cycle. However, one may generalize at least to the extent of saying that the major factors which ensure the maximum heat-fastness in a pigment are insolubility, thermal stability and lack of reactivity.

Migration properties

Migration of colour from polymers is related to the diffusion coefficient and causes two main defects. Contact bleed, or cross-staining occurs when a dyestuff or pigment is sufficiently soluble to diffuse from the coloured material into an uncoloured, or differently coloured material with which it may come into contact. For example, if a PVC cable covered with a migratory red pigment was covered by a white outer sheath of PVC, the latter would become pink as the red diffused into it. Similarly a red 'effect' thread in a fabric could stain adjacent threads.

The other defect arising from migration is bloom of the pigment to the surface of the polymer: this causes poor rub-fastness. In severe cases, photo-micrographs show crystals of colour growing out through the polymer surface like hair or wool growing through skin. It occurs in polymer systems where the pigment used has some solubility in the polymer at high processing temperatures, so that when the polymer is rapidly cooled to produce the shaped solid, the polymer is supersaturated with colour. Providing that the colourant molecules can diffuse through the polymer, crystallization or precipitation of the colour occurs slowly, and crystals grow at the surface where the concentration gradient is at a maximum.

Durability

Light-fastness and weathering-fastness are the principal considerations here: fading of the colour must not occur within the expected life of the article. For synthetic fibres and domestic articles, light-fastness data using modern accelerated-light-fading and established standard-test procedures are a reliable guide to performance. However, for outdoor applications such as building components, road signs, car bodies, or boats, much higher fastness to weathering is required, and accelerated-light-fastness data are not adequate. In addition, the durability of the polymer and the effect of the pigment on the resistance of the polymer to photodegradation must be considered. The fact that certain pigments are known to improve the life of polypropylene, for example, is taken into account in formulating colour recipes for bottle crates or marine ropes. The durability of one coloured polymer system can differ widely from that of another system even when the same pigments are used, and a colourant can only be given a reliable fastness rating when the polymer system is specified fully.

Miscellaneous properties required of pigments in synthetic polymers

In addition to the properties detailed above, a vast array of requirements arises from specific outlets and some of the more important ones are tabulated below.

Application	*Properties required*
Paints	Good hiding power, flow, gloss, and resistance to overspray.
Inks	Good flow, gloss, transparency for trichromatic printing.
Textiles	Good wash and dry-cleaning fastness, and fastness to other textile processes.
Plastics and Rubber	No effect on polymerization and curing systems. Non-bleeding from bottles and containers to their contents. Non-toxic for toys and food containers. No effect on electrical properties for insulation materials. Resistant to steam and sulphur cures.

Future prospects for pigments in synthetic polymers

The use of pigments is very old and well known: the point of interest is how the use has changed. Originally pigments seem to have been used for special occasions: decorated cave dwellings are relatively rare and the Pharaohs' tombs certainly were special. 'Old masters' and decorative architecture were sponsored by the church and secular princes of the Italian renaissance. Today synthetic chemistry has developed brighter pigments and in quantities and at prices suitable for everyday use over a wide range of articles.

The advent of the synthetic-polymer industry has provided outlets for pigments in synthetic paints and enamels, coloured plastics, and pigmented synthetic fibres. Reinforced plastics for boat construction and new uses in aviation have placed special demands on pigments. Recently more specialized developments, more in keeping with the artistic tradition, have occurred in printing inks: colour illustrations have become possible in high-quality books; and at the other end of this spectrum children's comics, glossy magazines, and newspaper illustrations, advertisements and 'colour supplements' have proliferated. Colour printing on synthetic packaging materials and for advertising has also developed rapidly.

Mass production of the motor car has provided several new outlets for pigments: industrial paints for motor cars, in quantities previously unthought of, pigmented leathercloth to replace leather in upholstery, and, of course, the mass pigmentation of rubber tyres. This glance at the present position suggests that major new markets for pigments will arise in the future either if *new* polymers find large-scale use or if polymers penetrate into new industries, e.g. the construction industry.

From the technological point of view, both existing and clearly-foreseeable markets trends are towards greater durability to weathering and higher temperatures. Other needs arise from the increasing trend to the replacement of natural by synthetic products, with retention of the traditional appearance. Patterns in vinyl panelling and flooring materials to replace wood, and thermosetting resin-bonded paper laminate for more durable working surfaces in kitchens and on tables, are examples. For the chemist, therefore, a vista is clear of increasing utilization of existing pigments, their modification to meet new markets, and the need for improved durability of pigments, achievable by intricate chemical synthesis within definite target costs.

3. Dyes for synthetic fibres

R. PRICE

THE chemistry and technology of the dyestuffs industry are probably more complex than those of any other section of the chemical industry; great number of products are manufactured by a diversity of processes. In fact, dyestuffs manufacturers are, by tradition, producers of fine chemicals on the large scale. The scale of activity of the dyestuffs industry is illustrated in Table 3.1, giving the quantities and value of synthetic organic dyestuffs produced in the major manufacturing countries during 1969.

The chemical industry can be divided into two major segments, the continuous-processing industry and the smaller-scale specialized chemicals industry. Dyestuffs manufacturers make relatively small quantities of large numbers of complex chemicals, both intermediates and dyestuffs, usually by batch processes. The complexity of the manufacture make the products expensive. Table 3.1 shows that the value of the dyestuffs averages £1500 per ton. The requirements of the dyestuffs manufacturers are for 'multi-purpose' (not specialized) plants so that a number of products can readily be made. The plant may be regarded as a scaled-up version of laboratory equipment; like such equipment it is assembled for a particular purpose, the manufacture of a specific chemical, and then used in a 'campaign' of several batches. A different array of equipment will probably be used for a different product in the next campaign.

To keep large stocks of dyestuffs which have a relatively high value is expensive—but so are frequent short plant campaigns expensive, since the plant is effectively lying idle during the necessary cleaning operations and rearrangement between campaigns. There is, therefore, a fine economic

TABLE 3.1

Production of dyestuffs for major manufacturing countries (O.E.C.D. *The Chemical Industry in Europe*, 1969/70, p. 125)

Country	Year	Production	
		Quantity (metric tons)	Value ($ million)
United States	1968	113 000	474·6
West Germany	1969	99 000	361·2
Japan	1969	66 000	134·3
United Kingdom	1969	43 000	139·1
Switzerland	1969	28 000	154·1

balance to be maintained for which the help of market forecasts and planning experts is essential in order to ensure the best deployment of resources.

The use of dyestuffs for colouring textile materials is long established. Dyes were originally natural colouring matters. With Perkin's synthesis of mauveine in 1856, however, a new era opened, and since that time the dyestuffs industry has developed rapidly. The industry does not deal directly with the ultimate user but serves dyers and fabric designers. It is governed by fashion and has the objective of providing the dyes that fulfil the requirements specified by customers. Among these requirements are the so-called 'fastness' properties of the dyestuffs. These are governed by the end-use of the dyed material. For example, long-lived furnishing materials, especially window curtains which spend much of their life fully exposed to sunlight, must resist fading or shade changes. The dyestuffs must have high light-fastness on the particular fibre used in the manufacture of the furnishing materials. High light-fastness would not be a requirement, however, in disposable items such as paper table napkins. Similarly dyes used on frequently laundered garments must possess good wash-fastness while the requirements for articles never in need of washing would be considerably less stringent in this respect. In addition to fastness under conditions likely to be encountered in use, dyes must be capable of withstanding a variety of different textile processes. Dyeing is possible at several stages, the one selected depending mainly on the type of material to be produced. For example, some fibres are dyed in the 'loose' state before they are spun into thread and woven into fabric. The dyes used for this purpose must be fast to the auxiliary agents and conditions employed in the subsequent processing. Some printed fabrics are subjected to a bleaching process, where the white areas of the print are cleaned up; the colours must remain unchanged through the bleaching process. In fact, nearly thirty different test methods can be used to determine the suitability of a dyestuff for a particular application. This presents the real challenge for the dyestuffs chemist: with our present knowledge it is impossible to generalize on the fastness properties of the various classes of dyestuff, especially since these vary considerably from fibre to fibre. For example, certain dyestuffs having very good light-fastness on nylon have very poor light-fastness on polypropylene. While this may not be too surprising, wide variations may occur within one general class of fibres—as from poly(ethylene terephthalate) to poly(1,4-cyclohexylenedimethylene terephthalate). To complicate matters even further, variations within a particular class of dyestuffs are not always in favour of the same fibre.

$$\left[\text{CO} - \left\langle \bigcirc \right\rangle - \text{CO}_2\text{CH}_2\text{CH}_2\text{O} \right]_n$$

Poly(ethylene terephthalate)

Poly(1,4-cyclohexylenedimethylene terephthalate)

Because of the wide range of fibres and textile materials, including blends, methods of application and processing, and end uses, the dyestuffs chemist has no lack of interesting problems to overcome. In fact, the history of the dyestuffs industry has been one of continuous change as advances are made and as new dyes are required for specific applications. The advent of synthetic fibres created a demand for whole new ranges of dyes, since those previously used on natural fibres were generally unsuitable for application on the new fibres. The allied dyeing industry is similarly complex because of the wide range of materials to be coloured, the variety of methods of application, and the diverse requirements to be met by the finished goods.

It is the objective of this chapter to present a picture—not in fine detail, for space will not permit this—of that part of the dyestuffs industry concerned with synthetic fibres, to highlight some of the probems which have been encountered by the dyestuffs chemist, and to indicate in which direction it is progressing. For more detailed reference, a bibliography is given on p. 85.

Synthetic dyestuffs

A dyestuff may be simply defined as a coloured substance that can be applied to and then interacts with a substrate, usually a textile fibre, normally from aqueous solution or suspension, so that the substrate acquires a coloured appearance. Dyes are coloured because they absorb electromagnetic radiation in the visible region of the spectrum. As a result of this, dyed textile materials reflect only a portion of the spectrum when they are viewed in white light. The colour seen is complementary to that absorbed. For example, if the dyestuff absorbs red light in the 13 000–16 500 cm^{-1} (605–750 nm) region of the visible spectrum, the dyed textile material appears greenish-blue.

Saturated organic compounds usually absorb electromagnetic radiation in the far ultraviolet, not in the visible region of the spectrum, and therefore appear colourless. The introduction of double bonds into organic compounds moves the absorption to lower frequency (longer wavelength) and this effect is enhanced by conjugation and by the presence of hetero-atoms such as nitrogen, oxygen, and sulphur in the conjugated system. By a suitable combination of these features in an organic molecule, its absorption maximum can be moved into the visible region of the spectrum so that the compound becomes coloured. The inclusion of certain substituents such as amino or hydroxyl groups in the molecule serves to magnify the effect. For example, biphenyl is colourless, azobenzene is weakly reddish-yellow, and 4-aminoazobenzene is quite strongly reddish-yellow. In this section we consider the chemistry, especially the synthesis, and the structural features of dyestuffs in relation to their properties.

Some thirty well-defined classes of chemical compounds find application as synthetic dyestuffs; the more important ones are listed in Table 3.2. These structures must be modified if desirable dyeing and fastness properties are to result for particular applications and this is the task of the dyestuffs chemist. Essentially this is a step-wise process. Starting from relatively simple and readily available raw materials, or primaries, one uses a variety of chemical

TABLE 3.2
Synthetic dyestuffs

Class	Typical dyestuff	Colour range
Azo		The whole of the visible spectrum
Anthraquinone		The whole of the visible spectrum Main commercial applications are bright reds and blues
Indigoid		Blue
Oxazine		Blues
Nitrodiphenylamine		Yellows
Methine		The whole of the visible spectrum

TABLE 3.2 (continued)

Class	Typical dyestuff	Colour range
Triarylmethane	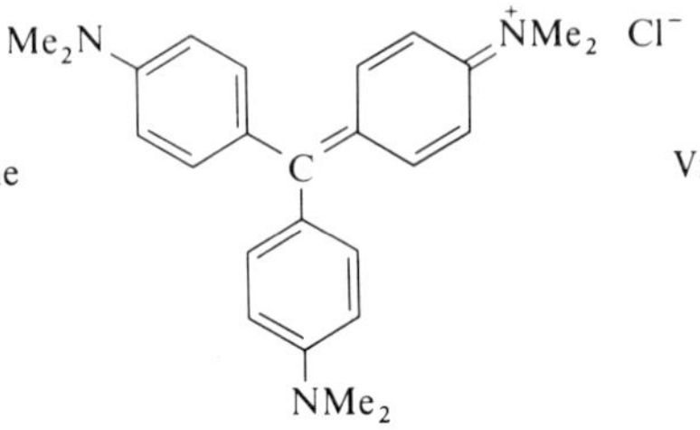	Violets, blues and greens
Quinphthalone	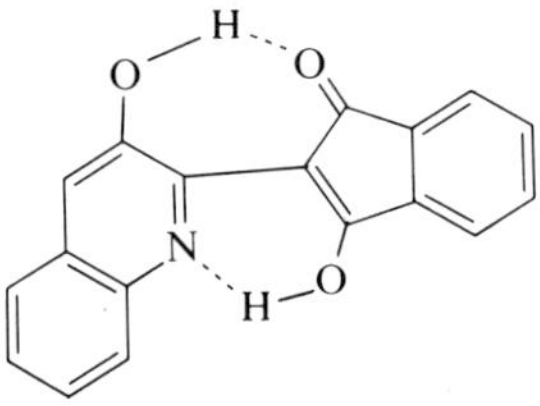	Yellows

reactions to produce suitable intermediates from which specific dyestuffs can be synthesized, often by a further complex sequence of reactions.

Azo dyestuffs

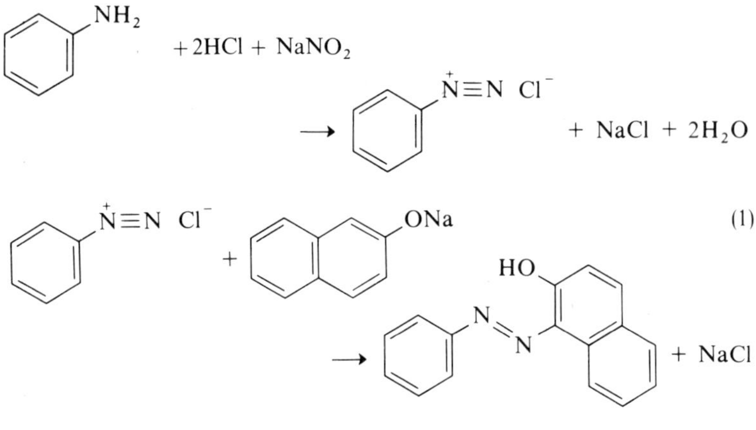

Without doubt the azo dyes represent the most extensive and widely used single class of dyestuffs, accounting for more than half of total manufacture. Almost invariably they are made by the well-known diazo-coupling reaction (Scheme 1). An amine is diazotized and the diazonium salt is coupled with a suitable coupling component, for example, a phenol, a naphthol, an aromatic amine, or a reactive methylene compound (such as an acetoacetarylamide or a pyrazolone). Those amines sufficiently basic to form salts with mineral acids are usually diazotized in aqueous medium; heterocyclic amines and homocyclic amines of low basicity, e.g. 5-nitro-2-aminothiazole, and 2-bromo-4,6-dinitroaniline, require non-aqueous media such as concentrated sulphuric acid, phosphoric acid, or certain organic acids in conjunction with nitrosyl sulphuric acid. In this reaction the attacking species NOX may be nitrous

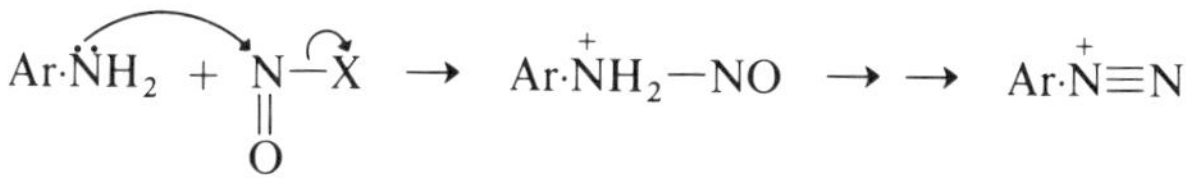

Ar·NH2 + N—X → Ar·NH2—NO → → Ar·N≡N
O

acid (X = OH), derived from sodium nitrite and hydrochloric acid, or a derivative of this acid such as nitrosyl chloride (X = Cl) or nitrosyl sulphuric acid (X = OSO_3H). The nitrosonium ion (NO^+) may also function as an attacking species.

Conditions for azo-coupling reactions depend upon the nature of the coupling component; phenols and naphthols couple via an ionized form; consequently, alkaline conditions favour the reaction:

Ar·N≡N
O
Ar·N=N H
O
Ar·N=N
HO

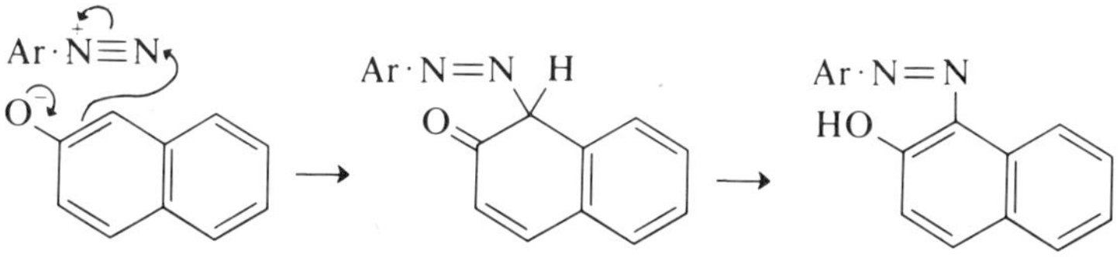

For coupling with aromatic amines, acidic conditions are best because the amine is thus solubilized as its salt. The use of too much acid must be avoided, however, since it is the free amine which takes part in the coupling reaction:

Me2N
N≡N·Ar
Me2N
H
N=N·Ar
Me2N
N
N·Ar

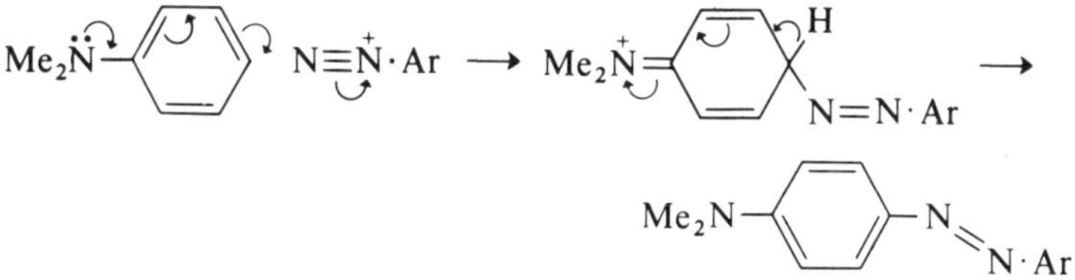

Aminonaphtholsulphonic acids, e.g. 1,8-aminonaphthol-3,6-disulphonic acid, are of particular interest since they can couple with diazonium compounds in either the 2-position under acidic conditions, or the 7-position under alkaline conditions. By coupling first under acidic and then under alkaline conditions it is possible to prepare bisazo dyestuffs from compounds of this type:

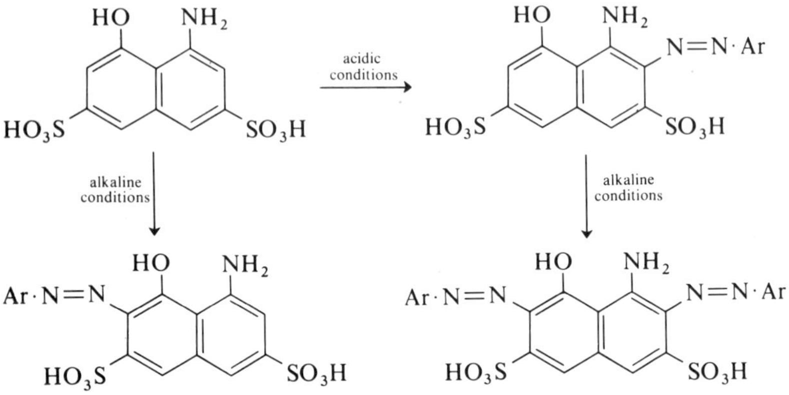

The very wide range of diazotizable amines and coupling components makes possible the synthesis of an enormous number of azo dyestuffs covering the whole visible spectrum. This number is increased even further by the possibility, already discussed above, of synthesizing bisazo-dyes containing two azo-groups, e.g. **3.1**, and dyes containing even greater numbers of azo-groups.

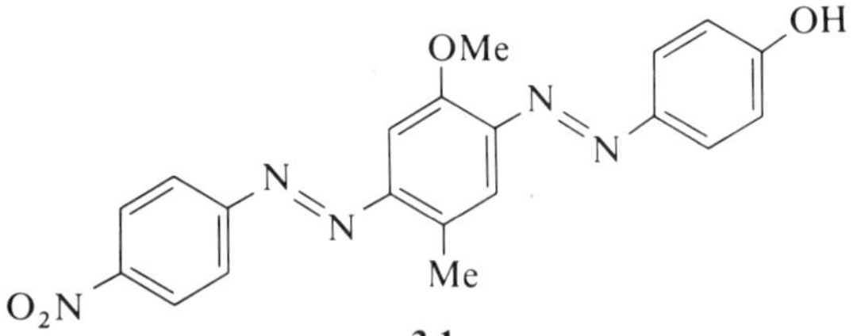

3.1

Monoazo dyes devoid of ionic groups are, in general, insoluble in water and are widely used as 'disperse' dyes for synthetic fibres. These are applied to the fibre as very fine dispersions in aqueous dyebaths and function by a process of dissolution of the dyestuff in the fibre. In the case of the very hydrophobic polyester fibre, penetration of the fibre by the dye is difficult at 100°C and it is important that the dyestuff should have small molecular size. This is a relatively easy requirement to meet in the yellow–orange–red shade-range with tertiary amine coupling-components and fairly simple diazo-components. Blue shades were not readily accessible however until heterocyclic diazo-components became available; to obtain suitable blue dyes before then, it was necessary to use diazo-components containing a multiplicity of electronegative groups. The effect on shade of progressively introducing electronegative groups into the diazo-component of monoazo dyes based upon tertiary-amine coupling-components is illustrated by formulae **3.2**, **3.3**, and **3.4**.

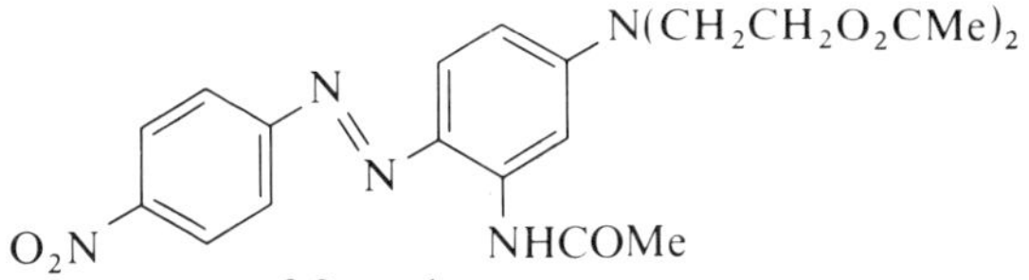

3.2, scarlet

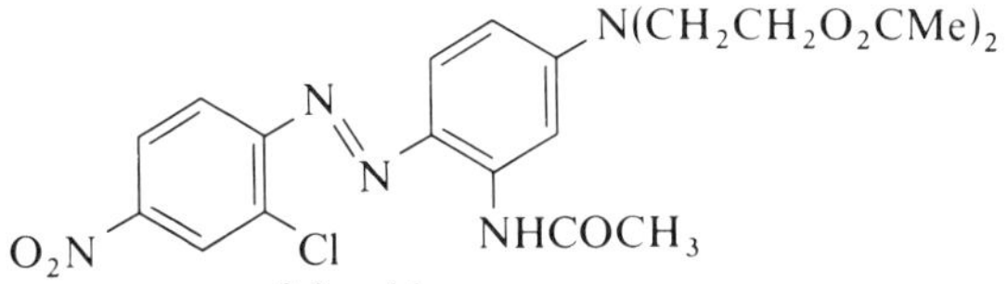

3.3, rubine

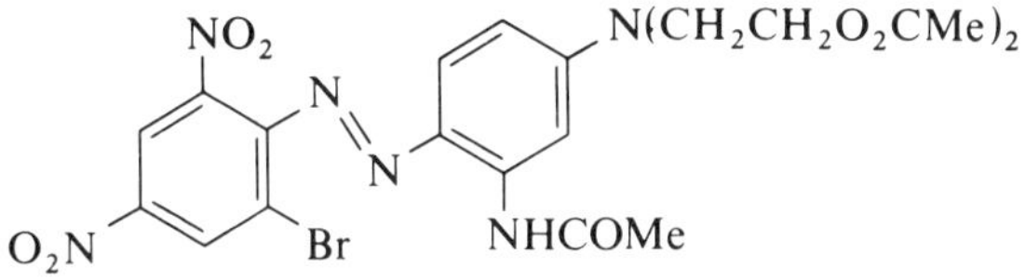

3.4, blue

Since 1950 increasing use has been made of the dramatic colour shift seen when heterocyclic diazo components are incorporated in relatively simple monoazo dyes. This is illustrated by compounds **3.5** and **3.6**: introduction of the thiazole system gives a blue dye. Use of this finding has provided

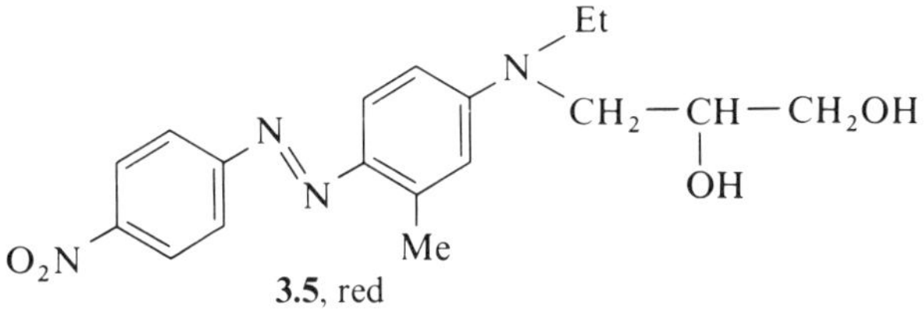

3.5, red

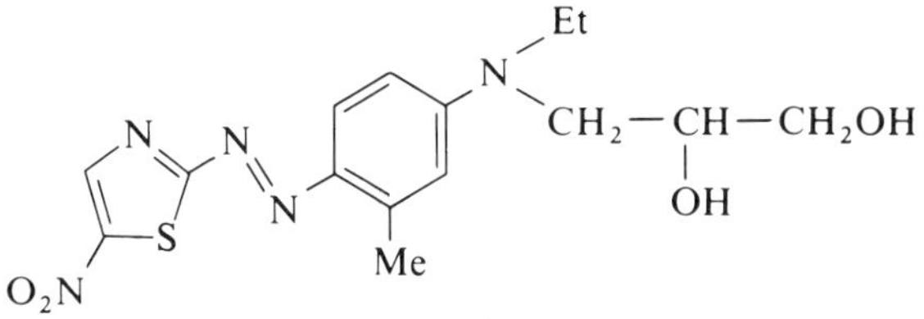

3.6, blue

disperse monoazo dyes, having high colour-intensity (extinction coefficients) and covering the whole of the visible spectrum. Suitable modification of other structural features, such as the nature of the alkyl groups attached to the tertiary nitrogen atom in the coupling component, has resulted in ranges of disperse dyes having good dyeing and fastness properties on polyester fibres.

The use of diazo- or coupling-components containing sulphonic acid groups results in anionic, water-soluble dyestuffs and monoazo dyes of this type, e.g. **3.7**, are used as 'acid' dyes for Nylon (see below). Water solubility

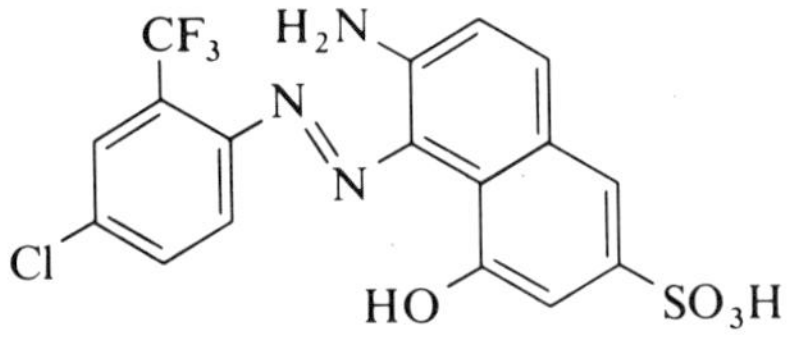

3.7, red

can also be conferred upon monoazo dyes by the introduction of quaternary ammonium groups. This is conveniently carried out by quaternization of water-insoluble monoazo dyes containing tertiary nitrogen atoms with, for example, dimethyl sulphate. 'Basic' dyes of this type, e.g. **3.8**, find wide application on polyacrylonitrile fibres (cf. p. 61).

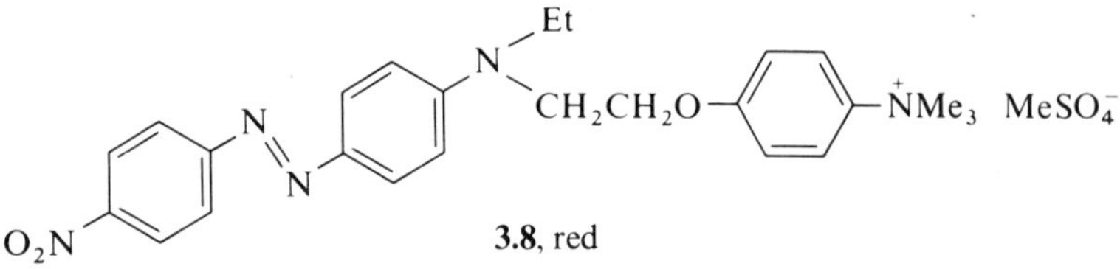

3.8, red

A selection of typical azo dyestuffs together with their applications is given in Table 3.3.

TABLE 3.3
Azo dyestuffs

Type	Typical dyestuff	Shade	Application
Anionic	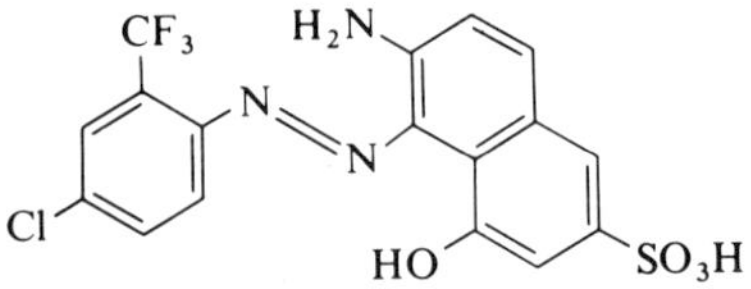	Red	Nylon

TABLE 3.3 (continued)

Type	Typical dyestuff	Shade	Application
Cationic		Red	Polyacrylonitrile
		Violet	
		Red	
		Blue	
Disperse		Orange	Polyester, Nylon, limited application on polyacrylonitrile
		Red	
		Blue	
		Blue	

Anthraquinone dyestuffs

With the exception of alizarin, which occurs naturally in madder and cochineal and has been used as a dyestuff since ancient times, anthraquinone dyestuffs date from the beginning of the twentieth century. They are relatively expensive but offer several advantages over other classes of dyes. A wide range of shades is available by suitable substitution of the anthraquinone nucleus, e.g. **3.9–3.12**. They also offer brightness of hue and very good fastness properties, particularly to light. They therefore rank among the most important classes of synthetic dyestuffs. The synthesis of most anthraquinone dyes starts from anthraquinone which itself is obtained by oxidation of anthracene or by cyclization of *o*-benzoylbenzoic acid from the Friedel–Crafts condensation of benzene and phthalic anhydride. Electrophilic substitution of the anthraquinone nucleus is difficult; halogenation and nitration cannot be used for the introduction of substituent groups. For example, the conditions necessary to effect nitration are so vigorous that the initially-formed 1-nitro derivative is heavily contaminated with the 1,5- and 1,8-dinitro derivatives.

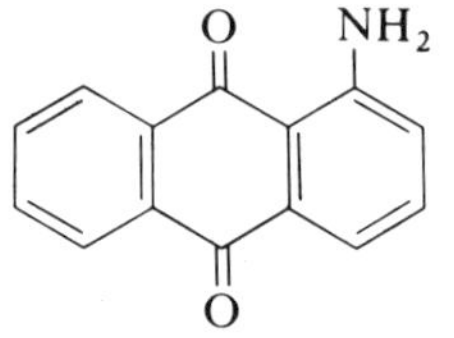

3.9, orange

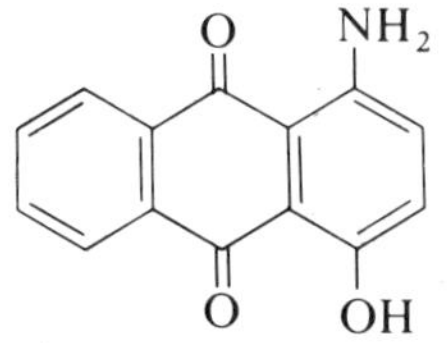

3.10, red

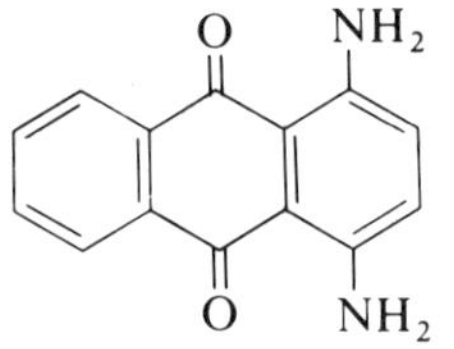

3.11, violet

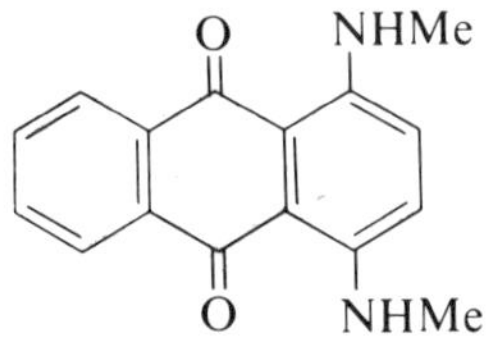

3.12, blue

Of these only the 1,8-dinitro compound can be easily purified; it is in fact used as a source of 1,8-diaminoanthraquinone. Sulphonation, however, can more readily be controlled. Normally it occurs in the β-positions of the anthraquinone nucleus, but in the presence of catalytic amounts of mercury, sulphonation occurs almost exclusively in the α-positions. By suitable choice of conditions, therefore, it is possible to synthesize a range of mono- and disulphonated anthraquinones. The sulphonic acid groups in these compounds are relatively labile and can be replaced by a variety of functional groups such as amino, substituted amino, hydroxyl, or chloro groups, which activate the anthraquinone nucleus towards electrophilic attack, so permitting the synthesis of a wide variety of substituted anthraquinones. A few typical examples are given in eqn (3.2).

$$(3.2)$$

Some important anthraquinone intermediates are manufactured directly by variations on the Friedel–Crafts method mentioned earlier. Thus reaction of phthalic anhydride with *p*-chlorophenol, followed by cyclization in concentrated sulphuric acid gives 1,4-dihydroxyanthraquinone (quinizarin) which, on condensation with methylamine, yields the blue disperse dyestuff **3.13** (eqn 3.3).

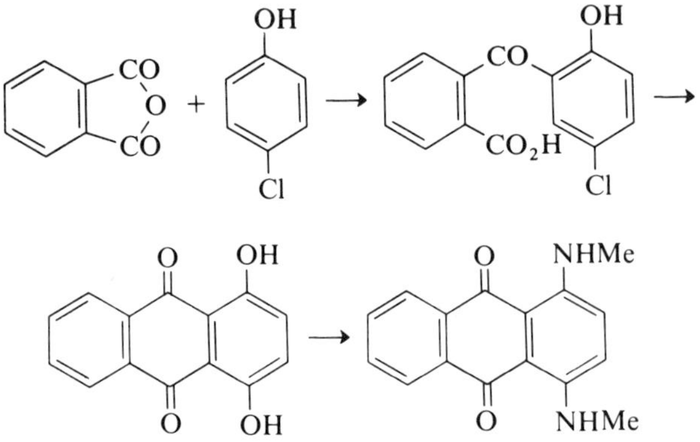

(3.3)

3.13

As with the azo dyes, very many different substituents have been incorporated in the anthraquinone nucleus to produce desired shades, and dyeing and fastness properties. For example dyes of the type **3.14** have been modified as in structure **3.15** in an attempt to overcome deficiencies in the sublimation fastness of **3.14** for certain applications.

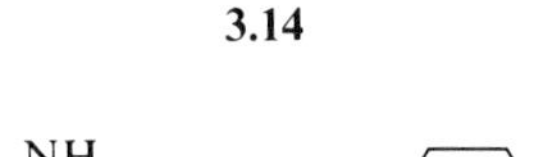

3.14

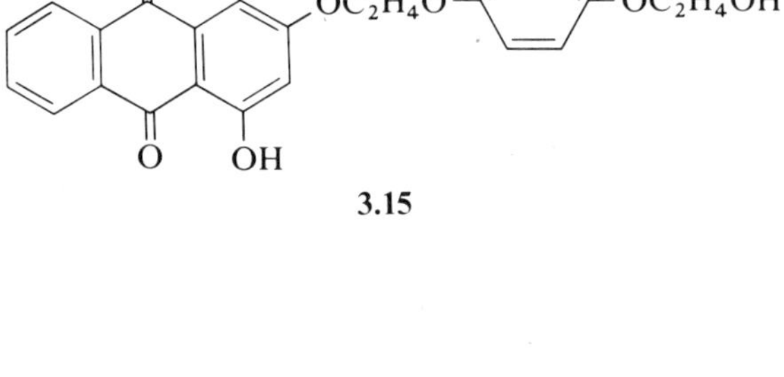

3.15

The introduction of sulphonic acid groups and quaternary amino groups leads to water-soluble 'acid' and 'basic' dyes, such as **3.16** and **3.17** respectively, suitable for application to polyamide and polyacrylonitrile fibres, respectively.

3.16, blue   **3.17**

The technology of dyeing synthetic fibres

General considerations

The prerequisites for a fibre to be dyeable are first, that it must be possible for the dyestuff to penetrate and second, that some interaction must take place between the dye and the fibre, so that the dye is not removed by subsequent processing, for example washing. Such interactions may be either physical or chemical. For example, mutual interactions between acidic groups in the dyestuff and basic groups in the fibre (or vice versa) can result in the formation of a salt-like link between the fibre and the dyestuff. Alternatively, when atoms in the dyestuff, in particular oxygen and nitrogen atoms, capable of donating electrons, are close to a suitable hydrogen atom in the fibre, hydrogen bonds are formed between the two centres. In certain cases, covalent bonds can be formed between the dye and the fibre by means of a chemical reaction between the two entities. In other cases, water-insoluble dyes may be formed within a fibre by the interaction of water-soluble precursors and these dyes are then physically retained. These principles had been well established before the synthetic fibres arrived on the scene in the 1940s and 1950s. Therefore, the dyeing of natural fibres will be discussed briefly in order to illustrate some of these effects and to highlight some of the new and different problems presented to the dyestuffs chemist by the advent of synthetic fibres.

The natural fibres traditionally used in textiles are cotton, wool, and silk. Cotton is hydrophilic. It consists of practically pure cellulose, a linear polymer containing recurring cellobiose units with a molecular weight of the order of 300 000–500 000. Because it is hydrophilic, the fibre is readily penetrated by aqueous solutions of dyestuffs but, unless some interaction occurs between the fibre and the dyestuff, the latter is easily removed again by washing. In the 'direct cotton dyes' it is believed that this interaction takes the form of hydrogen bonding between hydroxyl groups of cellobiose units in the fibre and amino, hydroxyl, or azo groups in the dye. Multiple interactions of this type between the fibre and the dyestuff and between individual dye molecules result in the formation of large aggregates of dye within the fibre pores, resistant to removal by washing.

In the vat-dyeing and azoic processes for the dyeing and printing of cotton, water-insoluble dyes are generated within the fibre. In the former process, a water-insoluble dyestuff is reduced with sodium dithionite in the presence of caustic soda to its water-soluble, leuco-form. Cotton is dyed with this

| Flavanthrone | Soluble 'leuco'-form |

solution and, on exposure to air, re-oxidation occurs with formation of the original insoluble dye within the fibre. In the azoic process, insoluble azo dyes are produced directly within the fibre by treating it with the individual component intermediates in a variety of ways. In its simplest form, the process consists of applying an azo-coupling component having cellulose affinity from an alkaline solution and then developing the impregnated fibre in a bath containing a diazonium compound. Insoluble particles of azo-dye are formed within the fibre and are caused to aggregate by a hot soaping treatment.

The most recent major development in dyes for cotton and other cellulosic fibres is the introduction of reactive dyes which form covalent bonds between the dye and the fibre by means of a chemical reaction; between, for example, a 2,4-dichloro-*s*-triazinylamino group in the dyestuff and a hydroxyl group in the cellulose.

Wool and silk are protein fibres. Wool consists mainly of keratin, built up from nineteen different amino acids, chiefly glutamic acid, cystine, leucine, arginine, and serine; the molecular weight is about 60 000. Some of the amino-acid residues in the molecule have pendant amino-groups and these, together with the terminal amino-groups, give the fibre a basic character. The fibre thus has an affinity for dyes containing sulphonic acid groups; it is the ionic attraction between this group and the basic groups in the wool that causes absorption by the fibre of the dyestuff from aqueous solution. However, the resulting link is easily broken and it cannot fully account for the good fastness to washing shown by many acid dyes. In these cases van der Waals forces must play an important part in the attachment of the dye to the fibre. Recently, reactive dyes for wool have been developed which owe their extremely good washing fastness properties to the formation of covalent bonds between acryloylamido-groups in the dyestuff and amino-groups in the wool.

$$Dye \cdot NH \cdot CO \cdot CH = CH_2 + NH_2 - wool \rightarrow$$
$$Dye \cdot NH \cdot CO \cdot CH_2 \cdot CH_2 \cdot NH - wool$$

Polyamides

The first truly-synthetic fibre to be produced on the manufacturing scale was Nylon, a polyamide obtained by condensation of the hexamethylene diamine salt of adipic acid. The average molecular weight was of the order of 12 000–20 000. *Nylon* is now regarded as a generic name, applied to a whole range of synthetic polyamides. Not surprisingly, in view of their chemical relationship to wool, which may be regarded as a complex natural polyamide, nylon polyamide fibres can be dyed with various types of acid dyestuffs; the dyestuff containing the sulphonic acid group diffuses through the fibre to form salt-like links with the terminal amino groups of the polymer. The acid dyes have better wet-fastness properties on nylon than on wool because of the more hydrophobic character of nylon but, in general, their light-fastness is somewhat poorer on nylon. However, some difficulties are encountered in the application of acid dyes to nylon. It is very difficult to manufacture nylon with uniform properties; variations occur in the fibre as a result of differences in physical and chemical structure arising respectively from different degrees of stretching and different terminal amino-group content. Consequently unless dyes with good 'levelling properties' (i.e. capable of uniformly covering these irregularities) are employed, unlevel or even streaky dyeing occurs.

Levelling of dyestuffs on the fibre depends upon migration of dyestuff molecules from one 'anchor site' to another and if this did not occur it would be impossible to achieve level dyeing. It is therefore essential that the bond formed between the dyestuff and the 'anchor site' is not so strong that migration cannot occur. However, it is precisely this bond which is responsible for the wash-fastness of the dyestuff on the fibre. The two properties of good levelling and good wash-fastness are therefore mutually incompatible. In practice it is necessary to arrive at a compromise with adequate levelling and wash-fastness for the particular end-use of the dyed fibre. For this reason, dyestuffs-manufacturers market ranges of acid dyes for nylon, classified according to levelling and wet-fastness properties. For example, the ICI Nylomine A range of acid dyes have good all-round properties, the Nylomine Acid B dyes have excellent levelling properties but in severe washing the shades lose depth, although there is little staining of adjacent white nylon, and the Nylomine Acid C dyes have outstanding wet-fastness properties but are inferior to the other classes in covering fibre irregularities.

Polyamide fibres can also be dyed with disperse dyes and these have the advantages of giving level dyeings and covering quite severe irregularities in the fibre. This, however, is achieved at the expense of wet-fastness, which is inferior to that achieved with the acid dyes. Disperse dyes are therefore used mainly in less heavy shades where wash-fastness is of less importance; for example, in the dyeing of ladies' hose and, sometimes, in conjunction with acid dyestuffs to overcome levelling problems. Typical disperse dyes for nylon are compounds **3.18** and **3.19**, both of which are based on azo-coupling components containing tertiary-amino groups. The tertiary-amino group is selected because it enables a wide range of shades to be obtained from relatively small dyestuff molecules easily capable of penetrating the fibre.

3.18, orange

3.19, red

Disperse azo-dyes, like **3.18** and **3.19**, have no sulphonic acid substituents. The sulphonic acid group not only confers water solubility which is undesirable in disperse dyes, but also contributes greatly to a dye's very desirable fastness to light. One of the major problems encountered with disperse azo-dyes, therefore, is the attainment of good light-fastness in molecules devoid of sulphonic acid groups. To produce this effect, non-ionic electronegative groups have to be incorporated in the molecule. The easiest such group to use is the nitro-group. In the case of nylon, however, there is the additional complication that in general azo-dyes containing nitro groups are degraded by the fibre. For this reason, many disperse azo-dyes for nylon are based upon diazo-components containing sulphonamido and alkyl-sulphonyl groups.

In 1959, ICI provided a solution to the problem of achieving level dyeings on nylon having good wash-fastness properties by introducing their Procinyl range of reactive disperse dyes. These dyes contain reactive halogen atoms and are applied to the nylon from weakly-acidic dyebaths. Under these conditions no reaction with the fibre occurs. When dyeing and levelling are complete, the dyebath is made alkaline to promote reaction between the dyestuff and the fibre. In this way, level dyeings having excellent wash-fastness are obtained.

Useful effects have been achieved by modifying polyamide fibres so as to change their affinities for dyestuffs. These modified fibres fall into two classes; those which by virtue of containing more basic centres than standard material have greater receptivity for acid dyes; and those that can be dyed with basic dyestuffs but have practically no affinity for acid dyes. Both types give level dyeings with disperse dyestuffs. Combinations of the several types of fibre with selected dyes provide patterned effects ranging from tone-in-tone to two-colour combinations. These find particular application in carpets and knitted goods.

Polyesters

Polyester fibres are based on poly(ethylene terephthalate), originally synthesized by Whinfield and Dickson. They have many properties very desirable in textile fibres: they are very hydrophobic and have high strength even when wet; they are resistant to attack by bacteria, moulds, moths, acids, and alkalies; their light-fastness is superior to that of nylon; and they have good heat-setting properties. Ironically these desirable properties present dyestuffs chemists with considerable problems: because the fibre is very hydrophobic and has a tightly-packed molecular structure devoid of reactive groups, it is unreceptive to dye molecules. In fact, polyester fibres of normal composition can be dyed only with certain disperse dyes; under normal dyeing conditions absorption of the dyestuff is very slow and only pale shades can be achieved. This results from the low diffusion coefficients of the usual disperse dyes in polyester at 100°C. In order to achieve reasonable depths of shade in economically acceptable times it is necessary to increase the rate of diffusion of the dyestuff inside the polyester fibre. This has been achieved in two ways; first, by increasing the dyeing temperature to about

130°C, which necessitates the use of pressure equipment; and second, by the use of 'carriers' at 100°C. Carriers are compounds which promote diffusion of the dyestuff within the fibre. Those in common use include biphenyl, 2-hydroxybiphenyl, 2,2'-dihydroxybiphenyl, and 1,2,4-trichlorobenzene. The use of carriers has certain disadvantages: they are rather expensive in use; difficulty is often experienced in completely removing the carrier from the dyed fibre; and in addition, the light fastness of the derived dyeings is sometimes impaired.

Polyester fibres can also be dyed by the du Pont 'Thermosol' process in which the fibre is first scoured and then impregnated with a disperse dye by use of a thickening agent and a suitable solvent. It is then dried and heated to 190–215°C for about one minute during which the applied dye is absorbed rapidly by the fibre. Any loose dyestuff is then removed and the fibre processed in the usual way.

Since diffusion of the dyes within the fibre is slow under conditions normally experienced in use, wet-fastness is good; yet the dyestuff is still free to migrate under certain conditions. This property is particularly noticeable when dyed fibres are subjected to heat treatments such as heat setting and pleating. Staining of adjacent white materials can then occur by diffusion of the dyestuff out of the fibre. For this reason, the choice of dyestuffs is critical for materials which are normally subjected to heat treatments.

The choice of dyestuffs is also affected by the fact that polyester fibre is often used in blends with cotton or viscose fibre. In the dyeing of such blends, disperse dyes are applied first to colour the polyester fibre; the cellulosic fibre is then dyed with reactive, direct or vat dyestuffs. It is important that the disperse dye used for the polyester should leave the cotton essentially unstained and the fastness properties of the dyeing thus unimpaired.

Polyacrylonitrile

The simplest polyacrylonitrile fibres are straight polymers of acrylonitrile and have little affinity for dyestuffs. Early polyacrylonitrile fibres were therefore very difficult to dye. However, various modified acrylic fibres are now produced by copolymerization of acrylonitrile with other monomers selected to confer affinity for dyestuffs on the resulting copolymers. The most common modified fibres contain sulphonic or carboxylic acid groups and can be dyed with cationic dyestuffs which become anchored within the fibre by salt-like bonds. While this results in very good wash-fastness in the dyeings, the very strength of the bonds creates difficulties in the application of such dyestuffs. The actual mechanism of dyeing involves, first, absorption of the dyestuff on the fibre surface by electrostatic forces. This is followed by dissolution of the dyestuff in the fibre and subsequent diffusion to acidic anchor sites. Once this has happened, migration from site to site takes place only very slowly and the achievement of level dyeings is difficult. It is, therefore, vital that the process of dyeing should achieve uniform distribution of the dyestuff on the fibre. This requires careful temperature control: although little uptake of dyestuff occurs below a critical temperature (usually this is above 85°C), above that temperature the rate of uptake of basic dyes on polyacrylonitrile increases rapidly. A further complication is that individual dyes vary considerably in their affinity for polyacrylonitrile. Those having a high affinity for the fibre show little tendency to migrate from anchor site to anchor site while those having a lower affinity for the fibre show a somewhat greater tendency to migrate. The latter dyestuffs can thus be displaced from anchor sites by the former. To avoid displacement effects or the blocking of sites by the dyestuff having greater affinity for them it is vital that any mixture of dyes applied to the fibre is mutually compatible.

Some solutions to these problems are now available. The levelling of cationic dyestuffs can be controlled to some extent by the use of so-called 'retarders'. These fall into two classes: anionic retarders which function by forming salts with the cationic dyes in the dyebath, thereby controlling their uptake by the fibre; and cationic retarders which function by competing for the anchor sites within the fibre. A disadvantage of the cationic retarders is that they do exert a blocking effect and thus reduce the depth of shade which can be obtained on a particular fibre.

In the early days of the fibre, selected, currently available cationic dyes were used. These gave dyeings having moderate light-fastness. The durability of acrylic fibres, however, demanded higher light-fastness than that available with these dyestuffs. As a result of the research effort devoted to this problem, new cationic dyes are now available which combine excellent light-fastness, brightness, and wash-fastness.

Selected disperse dyes can also be used for dyeing polyacrylonitrile; they have good levelling properties but tend to be deficient during wet treatment and steam pleating, and are therefore generally considered only suitable for pale shades.

The wide variety of chromophoric systems available to the dyestuff chemist makes an exhaustive treatment of this dyestuff chemistry impossible. Research on cationic dyes for polyacrylonitrile has, however, provided a number of interesting comparisons and structure–activity correlations, which can be summarized briefly. The main types of chromophoric systems used are those in the azo, methine, azacyanine, anthraquinone, triphenylmethane, and oxazine dyestuffs.

The triphenylmethane dyestuffs, e.g. Malachite Green **3.20**, are characterized by high tinctorial strength and very bright shades but suffer from only poor to moderate fastness to light when dyed on polyacrylonitrile. Very bright shades and high tinctorial strength are also characteristic of the methine dyes but, in general, most dyes of this class have only moderate light-fastness on polyacrylonitrile. Examples of dyestuffs of this type include **3.21** and

3.22. Azacyanines, e.g. **3.23**, yield very bright shades having high light-fastness on acrylic fibres but are restricted to yellows. The oxazines, e.g. **3.24**, also give bright, tinctorially strong shades, mostly blues. The light-fastness properties of these dyes on polyacrylonitrile vary considerably depending upon substituents in the molecule but are rarely better than moderate to good.

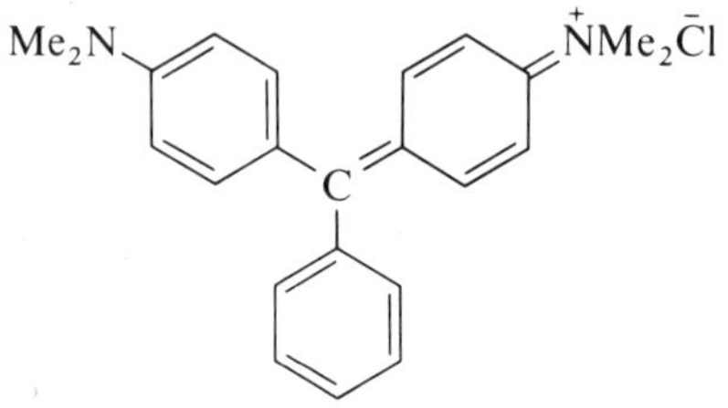

3.20

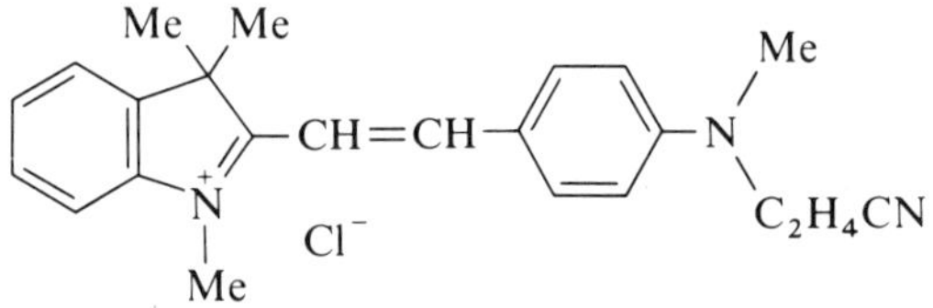

3.21, pink

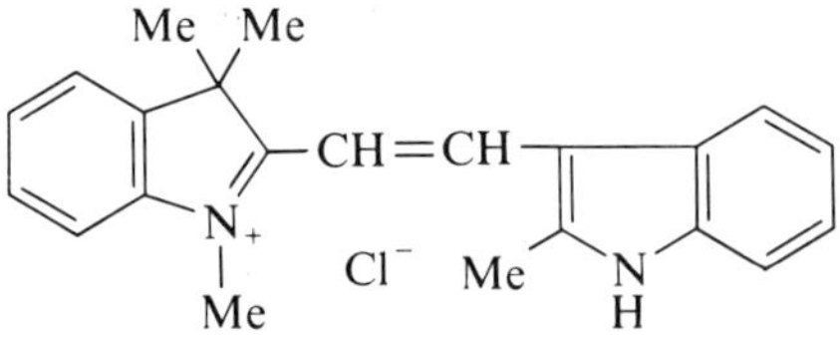

3.22, bright orange

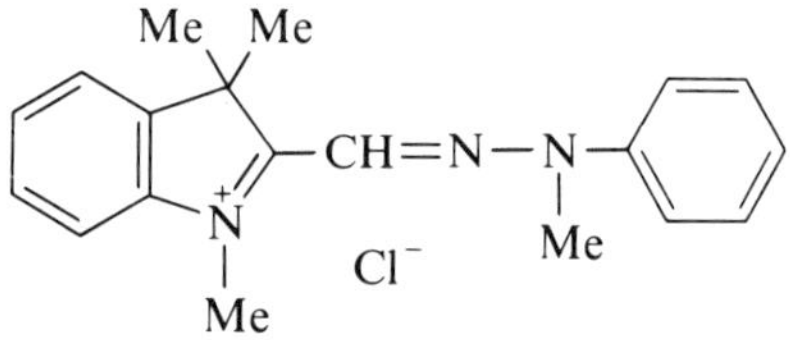

3.23, golden-yellow

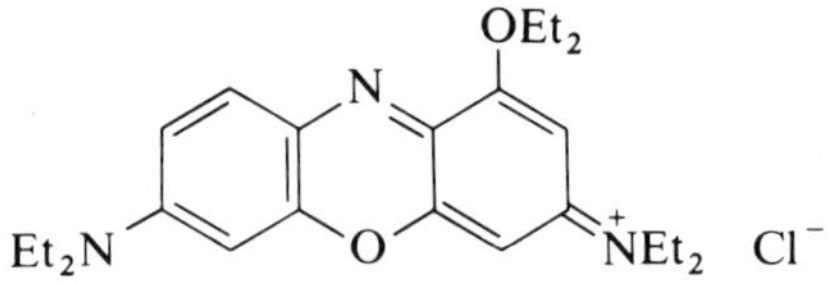

3.24, blue

The difficulties in obtaining blue dyes with good light-fastness can be overcome by using anthraquinone dyes. These have molar extinction coefficients (colour intensities) only about one half those of azo dyes and one quarter to one eighth those of triphenylmethane dyes but provide blues, for example **3.25**, with extremely good light-fastness on polyacrylonitrile.

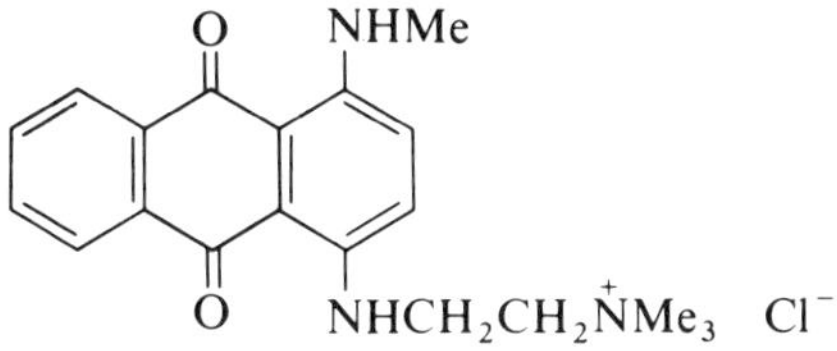

3.25

The most important single class of dyes for polyacrylonitrile, as with other fibres, is the azo series. Cationic azo dyes developed for polyacrylonitrile fall into two main classes; those, e.g. **3.26** and **3.27**, in which the cationic centre is remote from the chromophoric system, and those containing a cyclammonium (quaternized heterocyclic) system, e.g. **3.28** and **3.29**, in which

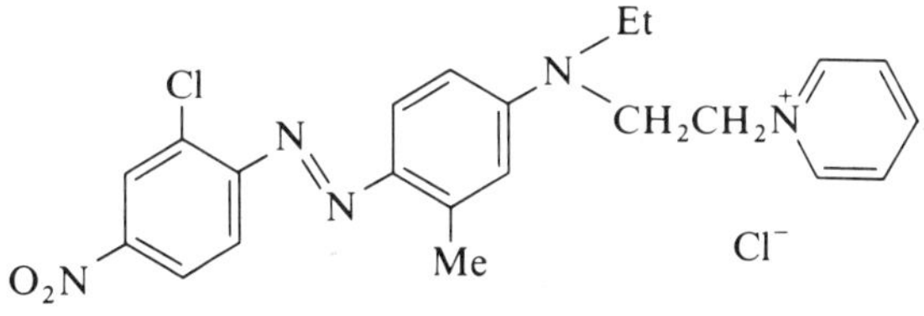

3.26, red

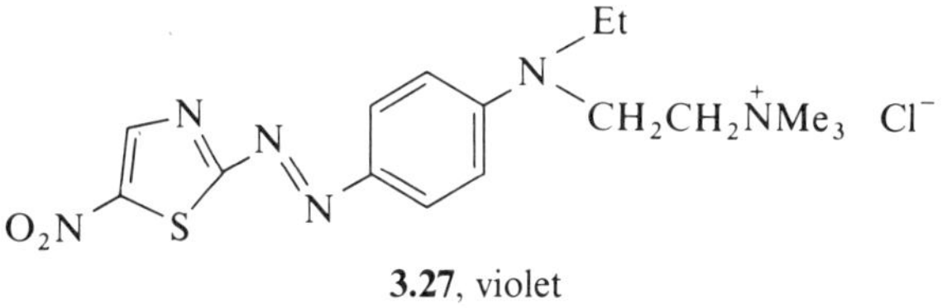

3.27, violet

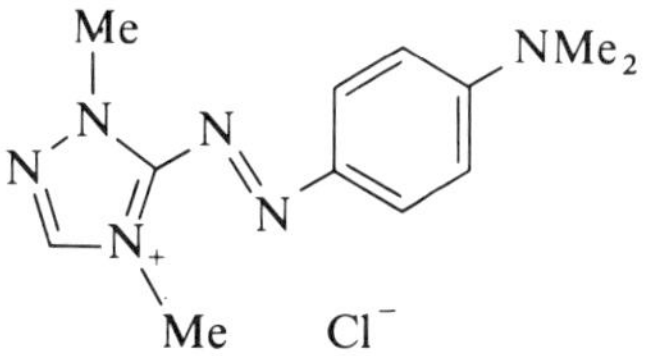

3.28, red 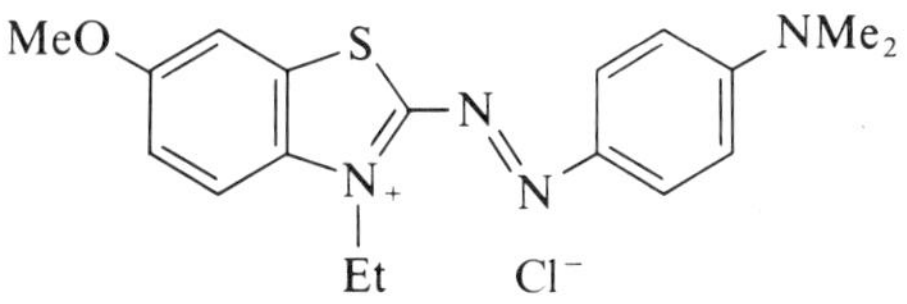   **3.29**, blue

the cationic centre is conjugated with the chromophoric system. Dyes of both classes give dyeings on polyacrylonitrile having good light and wash-fastness. The former are somewhat dull by comparison with triphenyl-methane or cyanine dyes, but the latter give very bright red and blue shades and have high tinctorial strength.

Thus, in general, dyes of the anthroquinone and azo classes have good fastness properties and those of the triphenylmethane, methine, azacyanine, and oxazine classes are characterized by intensity and brightness of shade but inferior fastness properties. The choice of dye is ultimately determined by a balance of physical properties and economic considerations with regard to the particular end use of the dyed fibre.

Polypropylene

Polyethylene finds a very large number of applications in the plastics industry but its low melting point makes it unsuitable for fibre production. Its higher-melting homologue polypropylene exists in a number of forms: in isotactic material the substituent methyl groups are all on the same side of the carbon-chain backbone; syndiotactic material has alternate methyl groups on opposite sides of the carbon-chain backbone; in atactic material the distribution of the methyl groups is more or less random (Fig. 3.1). The isotactic form is highly symmetrical and largely crystalline. It is therefore particularly suitable for spinning and drawing into fibres valued for their high tensile strength, chemical stability, low bulk density and low cost.

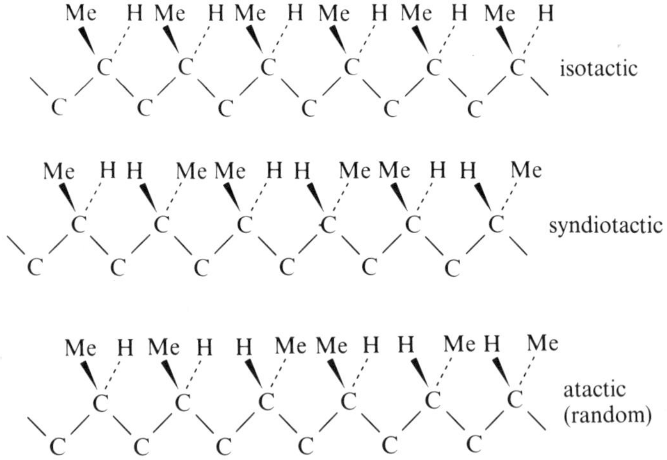

FIG. 3.1   Polypropylene stereochemistry.

Polypropylene fibres are, however, hydrophobic and impenetrable by water-soluble dyes. They moreover contain no functional groups capable of serving as anchor sites for dyestuff molecules. The oil-soluble Waxoline dyes penetrate the fibre when applied as aqueous dispersions and give medium-depth shades, but the resulting dyeings have a number of serious short-commings. The dyestuff tends to 'bloom' or migrate to the surface of the fibre on storage thus giving very poor rubbing-fastness. Fastness to light and to washing is fairly poor and dry-cleaning solvents remove the whole of the dyestuff from the fibre.

Thus polypropylene presented the dyestuffs technologist with a number of novel problems. Several solutions have been devised. Many of them are based on the physical incorporation of acidic or basic substances in the melt from which the fibre is spun. For example, molten polypropylene may be admixed with poly(vinylpyridine) or a copolymer of N-vinylpyrrolidone and dimethylaminoethyl methacrylate; the fibres can then be dyed with conventional acid dyes. Similarly the incorporation of acidic materials in the melt gives fibres which can be dyed with conventional basic dyes.

Various methods have also been devised for chemically modifying polypropylene fibres to render them receptive to dyestuffs. For example, chlorination and bromination of the fibre are claimed to confer affinity for basic dyes. Most of these treatments, however, suffer from the drawback of principally affecting the surface of the fibre, thus giving rise to 'ring' dyed or incompletely penetrated fibres.

An interesting method of achieving dyeability in polypropylene consists of incorporating in the melt a suitable metallic compound, for example, aluminium stearate or the nickel complex of an alkylated o,o'-dihydroxydiphenyl sulphide or sulphone. The latter compounds serve a dual function in stabilizing the fibre against degradation by ultraviolet light and providing anchor sites for dyestuffs capable of combining with metal ions when applied to the fibre as aqueous dispersions. A wide variety of such metallizable azodyestuffs has been claimed in the patent literature as useful for this purpose; other metallizable dyes, such as suitably-substituted anthraquinones, are also used. The most important classes of azo dyes for this application include tridentate types such as o-carboxy-o'-hydroxydiarylazo (e.g. **3.30**), o,o'-dihydroxydiarylazo (e.g. **3.31**) or o-hydroxyazo dyes containing a suitably-situated heterocyclic donor atom, usually nitrogen, e.g. **3.32**, and bidentate types derived from coupling components such as 8-hydroxyquinoline (e.g. **3.33**), salicylic acid, or salicylaldoxime.

Dyestuffs of these types penetrate and become anchored within the fibre by the formation of metal complexes (e.g. **3.34**) with the additives or even with the small amounts of residual metal salts which may be derived from the polymerization catalysts.

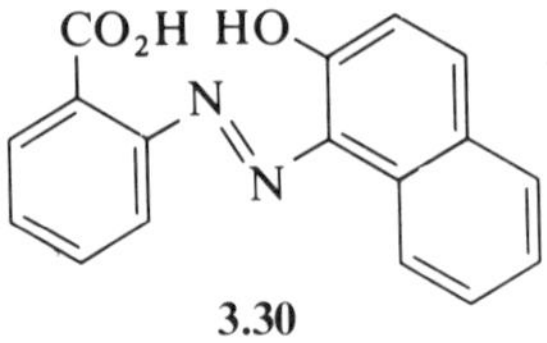

3.30

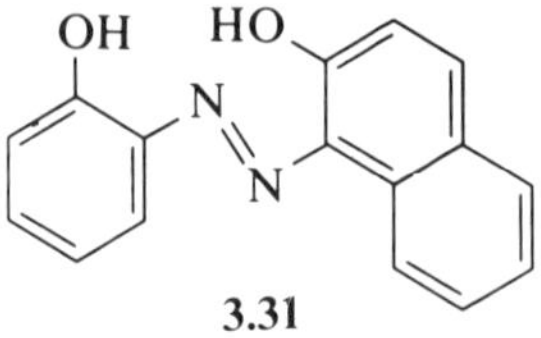

3.31

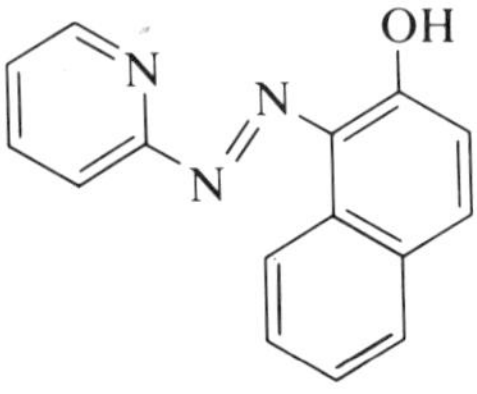

3.32

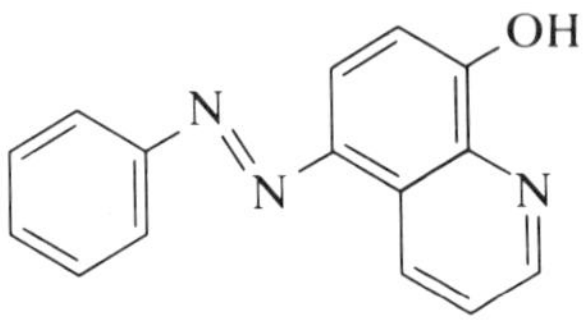

3.33

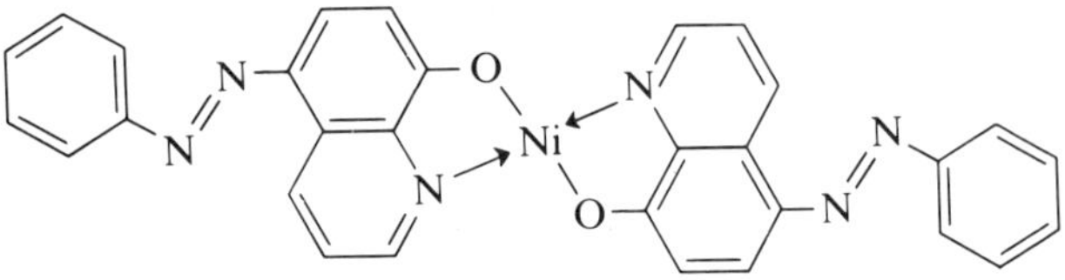

3.34

Future prospects

The synthetic dyestuffs industry has been in existence for over 100 years. Throughout that time intensive research effort has been devoted to improving its products and simplifying the work of the dyer. Even after all this time, there have recently been numerous innovations, both in the dyestuffs available and in the methods of applying them. Some of these were brought about by the advent of synthetic fibres and the increasing use of these fibres ensures that this will be a continuing process, with further progress towards brighter colours and more economic dyeing processes. The most recent forecast (P. Meunier, *Text. Chem. and Col.*, 1971, Dec., 37) of the prospects for the dyestuffs industry in the next decade relates to the United States (Table 3.4). The 1980 market, it is expected, will show a 70 per cent increase in value over the 1970 market. These figures are based on population growth, fibre consumption, and technological advances. Cationic, disperse, and acid dyes are expected to show the greatest growth as the market for acrylic, polyester, and polyamide textile materials expands (Table 3.5). Textured polyester in knitted and woven fabrics, blends, and carpets are identified as the major growth areas.

TABLE 3.4

Growth of dyestuffs sales in the United States
(P. Meunier, *Text. Chem. and Col.*, 1971, Dec., 37)

Year	Per capita consumption of dye ($ person^{-1} year^{-1})	U.S. population (millions)	Market for dyes ($ million)
1935	0·42	117·0	47
1950	1·20	150·6	180
1969	2·18	197·5	430
1975	2·50 (estimated)	220·6 (estimated)	550 (estimated)
1980	2·85 (estimated)	237·7 (estimated)	680 (estimated)

TABLE 3.5

Consumption of man-made fibres in the United States
(P. Meunier, *Text. Chem. and Col.*, 1971, Dec., 17)

Year	Acrylic (m lb)	Nylon (m lb)	Polyester (m lb)
1960	160	370	120
1969	590	1375	1270
1979	890	1760	3320

In the period 1968 to 1969 the share of the market enjoyed by disperse dyes increased by 17 per cent due to the growth of polyester and polyamide fibres, that of basic dyes increased by 16 per cent owing to the increased use of acrylic, modified acrylic ('modacrylic') and acid-modified polyester and polyamide fibres, and that of acid dyes increased by 9 per cent because of use on nylon carpets. In the same period the sales of vat dyestuffs fell by 5 per cent because of decreased use of cotton and blends.

The search for new chromophores and for new dyestuffs having technical and economic advantages over those currently available continues. A very desirable target is to produce only three primary dyestuffs for each fibre, suitable combinations of which would provide a complete spectrum of shades. Such dyestuffs would of course have to be mutually compatible in terms of dyeing and fastness properties and have virtually ideal absorption spectra; thus many problems must be overcome before this can be achieved. However, the very obvious benefits to both the manufacturer and the user ensure continued effort towards this target.

Other changes are in progress; for example the concept of applying dyes to fibres from a solution in an organic solvent or a mixture of solvents, though not new (having first been proposed some sixty years ago) was until recently of little interest. Now interest is increasing for a number of reasons, some economic and some resulting from current concern with the problems of environmental amenities. Organic solvents are already used for the scouring and finishing of textile materials and equipment used for these processes could be adapted for dyeing. This opens up the possibility of combining dyeing and finishing in a single unit operation with the added attractions of shorter dyeing cycles, lower heating costs and preservation of fabric dimensions and appearance. Together with the need to conserve water which is used in considerable quantities in conventional aqueous dyeing processes, and to avoid environmental pollution, these factors provide a powerful incentive to the development of solvent-dyeing methods.

An alternative approach lies in the intriguing prospect of dry dyeing in which the vapour pressure of the dyestuff is utilized. Direct transfer of dyestuff to the synthetic fibre into which it can diffuse, avoids the use of solvents, auxiliary chemicals, etc. Level dyeing of polyester packages has already been achieved by this method using the dyestuff **3.13**, which penetrates the fibre completely. In an extension of this process, pastes containing certain disperse dyes have been printed on paper. After drying, the paper is heated in close contact with synthetic fibres: sublimation of the dyestuff transfers the printed pattern to the fabric. The types of dyestuff which can be applied by these processes are at present very restricted and problems exist regarding the fastness properties of the derived dyeings. There is therefore room for improvement which the dyestuffs industry will set out to achieve over the next decade.

4. Synthetic fibres and fabrics: processing and finishing aids

J. W. BATTY

THE preceding chapter shows how the advent of synthetic fibres changed the products of the dyestuffs industry. New problems concerning dyestuffs arose, for example in attaching dyestuffs molecules to the new synthetic polymers, and in providing brighter and more effective dyes for them. These problems were solved by chemistry—indeed they illustrate how useful chemistry can be in modern industrial practice.

The new synthetic-fibre industry, however, posed many problems—the dyestuffs industry was concerned with but a small proportion of them. This chapter illustrates some of these problems, met by the fibre and textile industries, and the role of the chemist in dealing with them. The emphasis however is on the technology rather than the chemistry. To make the problems and, indeed, the changes in the industries more readily understandable, some background has first to be given.

Wholly-synthetic fibres are relatively new: the major inventions laying the foundations of the industry were made in the period between the two world wars. Three sorts of fibres account for most of the material made: polyamides (Nylon), polyesters [based on poly(ethylene terephthalate)], and acrylics (based on polyacrylonitrile). Polyalkene fibres are less important. From essentially nil in 1944, production of all wholly synthetic fibres grew to 0·4 million tons in 1958, to 4 million tons world-wide (worth >£2000 million) in 1970, and may be as high as 12 million tons by 1980. Thus more than 20 per cent by weight of all fibres used are now wholly synthetic, and this percentage is still growing.

The uses for these fibres are extremely varied, but reflect their special advantages, particularly their strength and elasticity. Polyamide and polyester fibres have exceptionally high strength and are used in ropes, tyre-cords, and carpets as well as in normal apparel. They are particularly suitable for hard-wearing clothing, both alone and as reinforcement for natural fibres in blends. Acrylics, because of their warm feel to the touch, are used mainly in household and apparel fabrics. Polyalkenes find use in ropes, twines, and in household fabrics (e.g. in deckchairs) where smoothness and cleanliness are an advantage.

These new materials did not penetrate the traditional market easily. The existing textile industry was based on natural fibres, especially wool and cotton, and had a major investment in plant to process these. It was prepared to accept innovation as long as it did not interfere too much with existing practice—especially when demand exceeded normal supply at the end of the war in 1945. The first objective therefore in the introduction of synthetic fibres was to minimize change, or rather the expense of changes, to the fibre user. Fibres had to be useable on existing machinery, thus avoiding purchases of novel machinery. Conventional textile machinery is designed for the major natural fibres—wool and cotton, both used in short lengths of staple fibre. For cotton the relatively short hairs are combed and then twisted together; a very tightly-twisted bundle of fibres is produced. Wool has a scaly surface and more natural crimp: in knitting wool the fibre is much less tightly twisted than in cotton. More tightly-twisted wool yarns provide 'worsteds'.

Synthetic fibres are produced by passing polymer through a spineret and bringing the very fine filaments together into one strong monofilament. The further handling of monofilaments however requires machinery different from wool- and cotton-processing machinery. Only a very small part of the older industry—that concerned with processing silk or rayon—had suitable machinery. To make synthetic fibres acceptable to the major part of the textile industry they had to be chopped into short lengths, in imitation of staple wool or cotton fibre. This step, demanded by the economics of the industry, was found to have a technical advantage, after all; the chopped synthetic fibre was easily blended with natural fibres and the resulting material was stronger than natural wool or cotton. These blends were particularly important in the early days when synthetic fibres were very expensive compared with natural fibres. The use of staple fibre thus greatly accelerated the penetration of the textile industry by synthetic fibres. Continuous-filament yarns were used only where their strength was important; for example nylon in stockings rapidly and completely displaced silk. Later examples of their use include nylon monofilaments in carpets subject to hard wear, and nylon and polyester monofilaments as replacement for rayon in tyre-cord for motor-car tyres.

As time progressed experience of advantages offered by synthetics was gained on the industrial scale; as machinery wore out it was replaced with these advantages in mind. Over the last decade or more, monofilaments have played an increasingly important role in the manufacture of fabrics, and to gain further advantages the manufacture of monofilaments has been improved. For example, economic considerations demand that machinery should process more and more material, i.e. an increase in the rate of throughput. This faster running imposes wear on machinery and on the filaments. Improved lubricants to overcome this difficulty are important textile aids.

Other aids used or sought to improve the processing of synthetic fibres are also discussed in this Chapter. There is an important interaction between the processing of synthetic, 'man-made' (cellulosic), and natural fibres: improvements made to one process are rapidly tested for the others. Many of these textile aids, therefore, are widely used not only for the synthetic fibres, but also for wool, cotton and blends. Obviously the chemist has his part to play in devising these processing aids, but he can only be effective if he complements his chemical knowledge by becoming fully familiar with textile technology.

Processing aids—lubricants and antistatics

The mechanical engineering processes by which polymer is converted into fabric are represented schematically in Fig. 4.1. The polymer is spun, drawn (to orient the fibre and provide maximum strength), and wound into

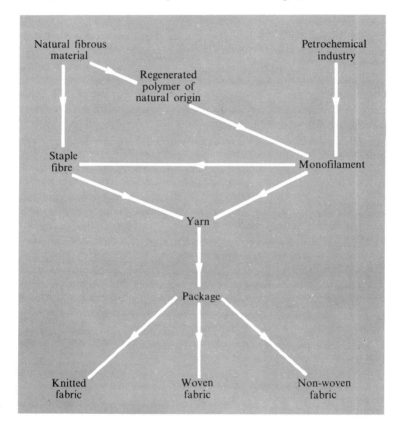

FIG. 4.1. Polymer to fabric: major stages.

suitable packages. The method of spinning depends on the polymer. Polyamides and polyesters are stable in the melt and are melt-spun. Cellulosic fibres and acrylics are not stable in the melt and are spun from solution. Good accounts of the techniques are available (Goodman 1967; Gordon Cook 1968). All stages of processing are improved by chemical processing aids. Some difficulties arise out of the fact that most synthetic fibres do not absorb water in a humid atmosphere as easily as do natural fibres. Consequently static charges tend to build up in processing. Fibres therefore either stick together or repel, instead of running easily through machinery. This is particularly noticeable in yarn production, when staple fibres are combed to orient the fibres in a uniform direction and are subsequently twisted together. The yarn is then drawn out and wound into some sort of package. Obviously friction between threads is important. The fibres and yarns are guided through the machinery by metal guides, and friction between the thread and the metal is also important. The requirements for a good processing aid are that it will control static charge production and lubricate the fibre so as to confer the correct amounts of friction to permit smooth processing. It should also be water-soluble, easily applied and removed from the fibre, non-corrosive to machinery, and without adverse effects on yarn properties. Preferably, the processing aid applied during manufacture should be effective also in the subsequent operations, but if an additional aid is applied it must be compatible with the original.

Both chemically and conceptually, simple aids are used. Materials absorbing water well (e.g. cellulosics and cotton) exhibit polar hydroxyl groups at the surface. It is assumed these are responsible for dissipating static charges. The chemical aids used imitate this effect: they are either hydrophilic or ionic and are applied as solutions to the thread during processing. For hydrophilic aids, it is hoped, moisture may be attracted to the surface, or a superficial film of moisture supported; the material would then behave as if it absorbed water well ('high moisture-regain'). For ionic aids, static charges may be dissipated by an ionic conduction process. As the raw materials and manufacturing techniques are now widely used, the processing aids are relatively cheap. A selection is listed in Table 4.1. The quantities used appear small in comparison with the weight of the fibre—generally in the range 0·1 to 1 per cent by weight: world consumption, however, is correspondingly large. To treat four million tons of fibre, tens of thousands of tons of processing aids are used annually.

Combinations of products from the different groups listed in Table 4.1 (p. 74) are used to obtain the balance of effects needed in a particular case. The mixing of anionic lubricants with cationic antistats must be avoided. The aids used are not usually pure chemical compounds—mixtures of closely-related natural products (e.g. fatty acids) or polymer fractions are cheaper and satisfactory in use. Purification is concerned with eliminating deleterious components. Chemically the agents include:

(1) For good lubrication, derivatives of fatty acids, esters, and amides from ethoxylated amines or polyamines.

(2) For good antistatic action, alkylphenol ethoxylates, alkyl phosphate salts, ethoxylated alcohols alone or as phosphate ester salts, fatty acid

ethoxylates, or fatty acid esters of polyoxyethylated compounds containing a high proportion of polyoxyethylene, particularly those derived from fatty acids of chain-length around C_{12}.

(3) For combined lubrication and antistatic action, fatty-acid ethoxylates are used, often with a C_{16} to C_{22} alkyl chain and upwards of 75 per cent polyoxyethylene content. Compounds with lower polyoxyethylene content are generally better lubricants but less efficient as antistats.

(4) Various types of surface active agents are used as bridging agents. Aromatic sulphonates can be used but may give trouble in later processing.

Lubricants are applied to all continuous-filament yarns during manufacture mostly as aqueous–mineral-oil emulsions with short-chain ethoxylates of aliphatic alcohols or alkylphenols. The aliphatic alcohol chain makes some contribution to lubricant properties. The short-chain ethoxylates have, in addition, some antistatic effect. If greater antistatic effect is needed ethoxylates of a higher polyoxyethylene content or an alkyl phosphate ester may be added, though this may cause some compatibility difficulties. Phosphate esters show other advantages, especially inhibition of metal corrosion. These are particularly valuable because powerfully surface-active agents, like the ethoxylates, may remove the surface films of grease on some machine parts, thus exposing them to corrosion. Actual compositions of the agents vary with the chemical nature of the fibre and the temperature and speed of processing.

Each method of processing presents interesting problems. Staple fibres, for example, are processed in one of several ways, depending on their chemical nature, differences in length and denier, and different amounts of crimp (which imparts fibre cohesion in the final package). The chemical nature of the fibre affects the inter-fibre friction and determines the amount of lubricant applied. Highly-crimped fibres are processed on cotton, woollen, and worsted textile machinery: longer, straighter fibres can be processed on machinery usually used with flax and jute. With cotton-type machinery, fine yarns are produced from short fibres parallelized at an early stage in the process. A balance has to be achieved between the ease with which the fibres slip over one another and fibre cohesion that, with added twist, controls the strength. Lack of lubrication gives too high cohesion in the fibres and results in uneven yarn. Excessive lubrication allows too much slip between fibres and results in weak yarns or breakages during processing. At best this is wasteful of reagents, at worst it stops the machinery. The antistat is also important: it is designed to prevent cohesion or repulsion (blowing out) of fibres which in bad cases can give almost a 'candyfloss' effect at a carding roller. Therefore, good antistatic agents with some lubricant action are generally used, i.e. an aliphatic acid or alcohol ethoxylate with C_{16} to C_{18} alkyl chains and a polyoxyethylene content of at least 60 per cent. Small amounts of other agents are added as required. Thus synthetic fibres or blends with a high

TABLE 4.1

Types of compound used in fibre processing

Type of compound	Typical formula	Remarks
(1) Good lubricants with some antistatic action		
Wax emulsions		Little antistatic effect from the dispersant
Fatty-acid esters	$C_{17}H_{35}CO_2C_4H_9$	Little or no antistatic effect
Stearic diethanolamide	$C_{17}H_{35}CON\begin{cases}CH_2CH_2OH\\CH_2CH_2OH\end{cases}$ $C_{17}H_{35}CO_2CH_2CH_2NHCH_2CH_2OH$	Commercial products are mixtures of these: proportions depend on conditions of preparation
Stearic acid ethanolaminoethylamides	$C_{17}H_{35}CONHCH_2CH_2NHCH_2CH_2OH$ $C_{17}H_{35}\!\!\underset{N}{\overset{N}{\bigsqcup}}\!\!N-CH_2CH_2OH$	Commercial products are mainly the first but prolonged heating or high reaction temperature causes ring closure
Stearic acid–polyethylene–polyamine	$C_{17}H_{35}CON\begin{cases}CH_2CH_2NH_2\\ [CH_2CH_2NH]_nCH_2CH_2\\ \qquad\qquad\qquad\qquad NH_2\end{cases}$ $n = 0 \text{ to } 3$	Commercial products are mixtures of various amides by reaction at several N-atoms. Ring closure as in previous compounds can occur. Usually applied as acetates
(2) Good antistats with little or no lubricating action		
Alkylphenol ethoxylates	$R\!-\!\!\!\bigcirc\!\!\!-[OCH_2CH_2]_nOH$	$R = C_8$ or C_9 branched chain, $n = 7$ to 15

Polyolethoxylates mono-fatty acid ester	$CH_2[OCH_2CH_2]_aOH$ $CH[OCH_2CH_2]_bOH$ $CH_2[OCH_2CH_2]_cO \cdot COR$	$R = C_7$ to C_{11} alkyl $a+b+c = 10$ to 30
Alkyl phosphate salts	$RO \cdot P$ with OH, O^-, $H_2\overset{+}{N}(CH_2CH_2OH)_2$	$R = C_7$ to C_{10} branched chain
Alkyl ethoxylate phosphates	$R(OCH_2CH_2)_nOP(O)(OH)_2$	$R = C_4$ to C_{15} usually branched. Used as salts or as free acids
(3) Good antistatic agents/Good lubricants		
Fatty acid ethoxylates	$RCO[OCH_2CH_2]_xOH$	$R = C_{15}$ or C_{17} fatty chain. $x = 7$–30. Commercial products are mixtures of monoester, diester, and polyethylene glycol
Castor oil ethoxylates	$CH_2O_2CC_{17}H_{32}[OCH_2CH_2]_aOH$ $CH \cdot O_2CC_{17}H_{32}[OCH_2CH_2]_bOH$ $CH_2 \cdot O_2CC_{17}H_{32}[OCH_2CH_2]_cOH$	$a+b+c = 5$–15. Higher levels of ethoxylation can be used but products are then solid
Fatty alcohol ethoxylates	$R[O \cdot CH_2CH_2]_xOH$	R is $C_{16}H_{33}$ or $C_{18}H_{35}$; x is usually 15–30
(4) Bridging agents		
Fatty alcohol ethoxylates	$R[OCH_2CH_2]_xOH$	R may be C_{16} or C_{18} saturated or unsaturated. Useful products for particular uses may have x from 2 to 30.
Alkyl phenol ethoxylates	$R{-}\langle C_6H_4 \rangle{-}[OCH_2CH_2]_xOH$	R is C_8 or C_9 branched chain; x is usually between 5 and 15.

Note. All commercial products are mixtures and compositions depend on raw materials used and method of preparation.

proportion of synthetics probably need an addition of a purely antistatic processing aid, whereas crimped fibres may benefit by addition of a lubricant only. Wool-type yarns need more lubrication and the lubricant may later have to be removed: it is therefore designed to wash out easily. In weaving and knitting, too, lubrication and static control are essential. Again blended processing aids are used, though there is also some use of starches or natural gums as sizes.

Considerable ingenuity has been shown by the yarn and fabric manufacturers in producing the fibre blends which work best on their machinery. It might therefore be thought that the auxiliary aids required would rapidly become well-defined and the market become stable. This is not so, for several reasons. Fibre processing has become increasingly capital-intensive and competitive internationally. Profit margins are low and manufacturers, in trying to raise them, increasingly seek special advantages; higher speed and greater ease of processing are obvious targets. With staple fibres a limit seems to have been reached in this respect. Future gains now appear most promising in processing of continuous-filament yarns which can be texturized at high speed and at above 200°C. Most present agents tend to decompose at these temperatures and cause not only discolouration of the fibres, but also deposits on ceramic guides used in the texturizing zone. These tend to break or weaken fibres. Excessive loss of agent through volatilization from the very thin film on the fibre surface can also lead to difficulties and variability in later operations. Some high-boiling liquid esters of fatty acids have proved effective lubricants but development of really heat-stable processing aids of all types is an important activity for the industry of the 1970s.

Hitherto, most aids have been biologically 'hard', i.e. not metabolized by micro-organisms; they are therefore possible pollutants if discharged to waste. Their use could be prohibited in due course. It seems inevitable that, before long, biodegradeable processing aids will be devised. This is, after all, what product development is about.

Chemicals improving properties of fabrics

As the use of the synthetic fibres has expanded and their commercial importance grown, their properties have been examined in great detail; in some respects they are equal or superior to natural fibres, but in other respects they may be inferior or perhaps merely different and therefore troublesome. When a new material is introduced even after preliminary test marketing, individual customers will evaluate it in use and note its deficiencies or differences from the traditional material first; the advantages will attract less and later comment. It is in obviating the disadvantages and thus overcoming initial customer resistance that 'finishing aids' find immediate use. The disadvantages particularly worrying to the fibremaker are the collection of static electricity on fabrics, the obviousness of deposited dirt (usually called 'soil'), and the combustibility of the products.

Controlling the build-up of electric charge

The build-up of electric charge is well-known in natural fibres: indeed in the classical school physics experiments, static electricity is generated by rubbing ebonite with silk cloth. The rubbing brings the surfaces into close contact and then separates them; it is not friction, but this separation which leaves electric charges on the separated surfaces and causes static charge to build up on fabrics. In fabrics made of natural fibres the static charge is troublesome only in *very* dry conditions, rare in the U.K. On synthetic fibres the build-up is more marked and shocks from carpets and furnishings are well known in the U.K., though still more unpleasant in countries with dry climates. The problem was well recognized in the early studies on synthetics. Indeed, it had to be solved in the processing of fibre and yarn (see above); in the fabric, however, it is impossible continuously to apply solutions to the material or to reduce the static generated by keeping the material damp.

In these early studies it was also realized that electrically-charged polymers attract dust and dirt particles and thus become soiled. The problems of 'antistatic' and 'antisoiling' were thus coupled and for a time confused. They are, in fact, separate problems. In the long run, it is highly desirable to cure both difficulties simultaneously; but to permit correlations between chemical structure and suppression or amelioration of undesirable activity, the two problems are best analysed and tackled individually. So far no really successful solution has emerged preventing the build-up of electrostatic charge and soiling. Some success was claimed initially with a polymerized quaternary salt (methacrylatoethyldiethylmethylammonium methosulphate, **4.1**) purely as an antistat. Moreover when the fabrics containing this polymer

$$CH_2{:}CHMeCO_2CH_2CH_2\overset{+}{N}R_3, \ MeSO_4^- \quad \textbf{4.1}, R = alkyl$$

were washed the methosulphate ion was replaced by either stearate or alkyl-benzenesulphonate, depending on whether a soap or a synthetic detergent was used: the resultant polymer salts were less soluble than the original. This agent was not easily removed by washing therefore and active for quite long periods ('durable'). However, these modified salts were less effective than the methosulphate as antistats and *in the wash* the charge-ionization which made them useful, caused them to attract particles of dirt (from other soiled garments in the wash) so that the garment often emerged from the wash dirtier than it was before. This defect ('wet soiling') is common to most quaternary ammonium compounds, and killed the use of this type of compound.

Attention therefore turned to polyoxyethylene compounds, particularly since aliphatic alcohol and acid derivatives containing these groups are effective non-durable antistats in fibre processing (Table 4.1). Success is still elusive, though many formulations containing polyoxyethylene groups have been tried. Usually the formulations also incorporated a polymerizable resin-mixture; to get durability of antistatic action the polymerization was completed on the fabric. Some useful effect was obtained and without too much soiling; but the fibres were stiffened and durability was generally poor. The advantage was not found to be sufficient in practice to compensate for these defects. Better results were obtained with polymers similar to the substrate and containing polyoxyethylene groups, but again durability was insufficient: with these products antisoil properties were reasonable and the soil release effect to repeated washing was better than the antistatic effect. The antistatic effect however was not as great as is required in practice. Since the compounds are designed to be most effective on substrates chemically similar to themselves, their use in blends is difficult and their efficacy then further reduced.

The present position is that several agents and treatments are on the market for particular fibres (Table 4.2), but there is no complete cheap solution yet to the durable antistatic problem in apparel fibres. Furnishing fabrics present an easier target, as wearer factors such as stiffness are less important, and the materials are only rarely laundered.

The overall antistatic problem may be solved only when cheap modification to the fibre polymers proves possible, so that materials with built-in anti-static properties can be manufactured. This is very expensive at present for polyamides, polyesters, and polyalkenes. It is somewhat easier to modify the structure of acrylic fibres and some of the so-called modacrylics have been designed to have improved properties, such as a better ability to dissipate static charge. Other techniques also offer some hope. For example, metal fibres have been introduced into polyamide fabrics to give a discharge route for static charges produced; this, though expensive, is at least partly success-ful.

Antisoiling

A great deal of attention has been paid to soiling in polyamide and polyester fibres, which are widely used in apparel, carpets, and household fabrics. There are distinct targets here since with apparel fabrics the main considera-tion is removal of greasy soil and stains by washing, whilst on carpets and household furnishings the main emphasis is on hiding the soil and removing dry particulate soil.

Success in providing for apparel fabrics either fibres or finishing treatments which prevent soiling in wear has been limited. Products are based on two hypotheses: (1) that to coat fabrics with 'low-energy' surfaces will avoid adherence of soil; therefore fluorochemicals, which have low surface tension and are not readily wetted at all even by mineral and vegetable oils, are used, and (2) that application of hydrophilic polymeric finishes to the fabric will repel oils and greasy soils and ensure ready wetting of the fabric surface by aqueous detergent. Soil-release finishes have been used mainly in the

TABLE 4.2

Types of compounds used as durable antistats

Type of compound	Typical formula	Remarks
Quaternary acrylates	$\left[\begin{array}{c} \text{Me} \\ \mid \\ \text{Et}_2\overset{+}{\text{N}}\text{CH}_2\text{CH}_2\text{O}_2\text{CC}=\text{CH}_2 \\ \mid \\ \text{Me} \quad \text{MeSO}_4^- \end{array}\right]_n$	Polymerized in situ
Acrylates containing polyoxyethylene groups	$\left[\begin{array}{c} \text{Me} \\ \mid \\ \text{CH}_2=\text{CCO}[\text{OCH}_2\text{CH}_2]_x\text{OH} \end{array}\right]_n$	Polymerized in situ alone or copolymerized with acrylic acid or alkyl acrylic ester; $x = 3$–10 but may be larger in copolymers
Ethoxylated Nylon 6	$\left[\begin{array}{c} \text{N[CH}_2]_5\text{CO} \\ \mid \\ [\text{CH}_2\text{CH}_2\text{O}]_x\text{H} \end{array}\right]_n$	Added to polymer melt and spun. $x = 10$–15
Ethoxylated castor oil	$\begin{array}{l} \text{CH}_2-\text{O}_2\text{C}-\text{C}_{17}\text{H}_{32}[\text{O·CH}_2\text{·CH}_2]_a\text{OH} \\ \text{CH}-\text{O}_2\text{C}-\text{C}_{17}\text{H}_{32}[\text{O·CH}_2\text{·CH}_2]_b\text{OH} \\ \text{CH}_2-\text{O}_2\text{C}-\text{C}_{17}\text{H}_{32}[\text{O·CH}_2\text{·CH}_2]_c\text{OH} \end{array}$	Added to polymer melt and spun; $a+b+c \sim 200$

USA for goods made of cotton–polyester blends. The products are of several types (cf. Table 4.3).

Fluorochemicals show good stain-release properties but no outstanding durability to laundering. Their price is high, but until an equally effective agent of lower cost is developed they will find wider application. Their competitors, acrylic emulsions, stiffen the fabric, in the quantities required to be effective. They show poor abrasion resistance and affect the shade of dyed fabrics. They have not made a major impact on the market. Polyesters similar in structure to the polyester fibre, but containing hydrophilic groups, particularly polyoxyethylene, give a reasonable degree of soil- and stain-release but are inferior to the fluorochemical and acrylic agents in release of oily soil. Ethoxylated Nylon-6 shows soil release properties on Nylon, but is insufficiently durable to washing. Improvements in all these types of agents are being made but no chemical type has yet appeared which really satisfies the textile technologist by providing durable soil release on more than one synthetic fibre.

Where polymers containing hydrophilic groups are effective on particular chemical substrates, e.g. polyamides or polyesters, built-in soil release properties are being sought by adding the polymeric agent to the fibre polymer before spinning. This procedure should ensure better wash durability than do applied finishes, but is still in the development stage.

On carpets and household furnishing fabrics, one line of approach has been the so-called 'white-soil' method. The assumption is that particulate soil becomes lodged in interstices in the fibre and fabric and is then difficult to remove; filling up these interstices would facilitate removal of the soil. This ingenious method has not had much success. Another has been the so-called 'soil-hiding' method. Soiling becomes visible because the light transmission through and reflections from, the fibres have altered; increasing the opacity should diminish the transmission and reflection and make detection of soil more difficult. The technique is already known; decreased lustre in fibres may be obtained by incorporating inorganic compounds. Methods have been developed of incorporating, at the fibre-spinning stage, fibre-immiscible polyoxyethylene compounds of high molecular weight. They are largely removed during subsequent processing (scouring, dyeing) leaving 'microcavity yarns'. The effect, so far limited to polyamides, is to increase the opacity of the fibre and hide the soil. Microcavity yarns are now offered by various manufacturers. Such treatments modify the fibres at the spinning stage and are expensive, but the effect is sufficiently desirable to make the cost tolerable at present. No doubt cheaper techniques of fibre-treatment will displace these products.

Fabrics with better non-flam properties

Synthetic fibres in apparel and upholstery goods compete with natural fibres. All organic fibres constitute a fire risk, and the less the risk, the greater

TABLE 4.3

Types of compounds used in soil release

Type of compound	Typical formula	Remarks
Perfluoroalkyl acrylates	$[C_8F_{17} \cdot O \cdot CO \cdot CH = CH_2]_n$	Polymerized in situ
Ethoxylated nylon-6	$\left[\begin{array}{c} -N[\cdot CH_2 \cdot]_5 CO- \\ [CH_2 CH_2 O]_x H \end{array} \right]_n$	$x = 8\text{-}15$
$-$Polyester containing polyoxyethylene groups $-[OCH_2CH_2CO_2$ CO-$[OCH_2CH_2]_x$ O_2C $CO]_n$		$x = 10\text{-}40$ units. About $\frac{1}{4}$ of the glycol is replaced

the competive edge. The risk is real; not only has legislation for 'non-flam' standards been widely demanded, but in some countries standards have already been defined. Important changes are happening or needed in high-risk situations such as children's and certain other clothing, and upholstery and furnishings particularly in hospitals, hotels, and aeroplanes. Both the actual degree of risk and methods of eliminating it are studied in detail. Until recently synthetic fibres were not generally considered very dangerous. Some of the polymers are thermoplastic and on ignition melt away from the flame; there is the problem of hot, molten or plastic, polymer adhering to the skin and causing burns, but since the polymer melts and falls away from the flame it seemed at least unlikely to propagate a serious fire. Unfortunately, in blends, the natural, e.g. cellulosic, fibres form a carbonaceous grid on combustion, on which synthetic fibre polymers burn quite well. Analogously, Nylon–glass-fibre blend is more flammable than nylon; and polyacrylonitrile fibre (not modified) is partly converted on heating into a solid polymer which is non-inflammable but also non-melting, and acts as a support on which polyacrylonitrile will burn instead of melting away from the flame.

Because of their wide use in carpeting and upholstery, acrylic fibres have received special attention. They have been modified with copolymers or blends of polymers, by use of applied finishes, and by additions to the fibre before spinning. Copolymers with vinyl chloride show some improvement in 'non-flam' properties, but not enough; non-flammability needs about 10 per cent of bromine or 25 per cent of chlorine in the fabric polymer and at such levels fibre properties are seriously impaired. Applied finishes have included ammonium bromide, phosphate, and sulphide applied with a urea–formaldehyde resin but again a high level of application is needed for reasonable effect; the applied resin stiffens the fabric and causes other defects.

Incorporation of suitable compounds ('additives') (Table 4.4) in the fibre before spinning has probably had most success; they are generally halogen-containing phosphates such as tris(dibromopropyl) phosphate, polyepi-chlorohydrin with calcium phosphate, or halogen-containing polymeric phosphonates. The usual type of flame-retardants seems less effective for poly-amides. In applications such as upholstery and automobile fabrics useful

TABLE 4.4

Types of compounds used in flame proofing

Type of compound	Typical formula	Remarks
Halogenoalkyl phosphates	$O{=}P(OCHBrCH_2Br)_3$	Included in polymer melt and spun
Phosphonium salts	$(HOCH_2)_4P^+, Cl^-$	Effective on mixtures with natural fibres

flame-resistance can be obtained by incorporating a flame-retardant in the backing to the fabric and this method is increasingly used. But further improvements in non-flammability are still being widely sought.

Similar considerations apply to polyester and polypropylene fibres. In upholstery, flame-retardants in the fibre backing are effective, but polyesters particularly find use in waddings, furnishings, blankets, and apparel, and very high non-flam standards are required for these.

Miscellaneous effects, including improvement of handle

A number of other chemicals are used with synthetic fibres to give necessary or desirable effects. For example compounds are added to stabilize the polymers against oxidation in preparation, in high-temperature manipulation during processing, and on extended exposure in use, or against the degrading effects of sunlight or u.v. light, or discolouration by long exposure to ordinary light. These compounds are also used in bulk and moulded polymers and are discussed in Chapter 1.

Some agents (Table 4.5) have been used to improve the handle of finished fabrics, particularly polyacrylics. Usually these take advantage of the characteristic feel imparted by long-chain alkyl groups, derived from fatty acids. Such agents are also used with natural-fibre fabrics, but their durability to washing is not good. More durable compounds which are more easily applied are being developed.

TABLE 4.5

Types of compounds used to improve handle

Type of compound	Typical formula	Remarks
Stearic acid–ethanolamino-ethylamide	$C_{17}H_{35}CONHCH_2CH_2NHCH_2CH_2OH$ or $C_{17}H_{35}-\langle$ imidazoline ring with two N $\rangle$	Applied with various binding agents to improve durability
Stearic acid–polythene polyamide	$C_{17}H_{35}CON\big\langle{}^{CH_2CH_2NH_2}_{[CH_2CH_2NH]_n \cdot CH_2CH_2NH_2}$ CH_2CH_2OH	Applied with various binding agents to improve durability

In this chapter the emphasis has been on the role and development of auxiliary chemicals and on their importance in increasing the acceptability of synthetic fibres and fabrics. But the major expansion of the synthetic fibre industry has been due to improved quality and cost reductions by better processing and making up of synthetic fibres. Possibilities of further advances by these methods are far from exhausted. A good account of the fibre industry in this respect is given by Moncrieff (1970) who also includes data on fibre usages and price movements.

The future for special-effects chemicals

The feeling in the textile industry has always been that better fibres can and should be developed, free from defects in the existing fibres, and retaining their valuable and attractive properties. Indeed, some success has been achieved in producing more easily-dyed polyamides with increased amino end-groups which have an affinity for acidic dyes by salt-formation. If these objectives are achieved, the need for many auxiliary chemicals may disappear. How likely is this? Over the last decade or so, economies of scale in the production of the existing fibres have begun to be achieved; the unit cost of fibres has dropped markedly as plant sizes have increased. To compete with the established fibres a new fibre made from a new polymer will have to have very remarkable properties and call for an immense investment (exceeding £50 million). Thus a major new fibre does not seem likely to displace existing products.

It is important therefore to evaluate fully existing processing and finishing aids and to improve them. From what has been said in this chapter it is clear that the finishing methods traditionally used are only partly successful when applied to fibres, synthetic or natural. There are one or two useful leads to further advances. Mixtures of polymers or additions to the polymers at the spinning stage may succeed in some cases, as already exemplified in 'micro-cavity fibres' and with flame-proofing additives. Composite fibres have been made having a central core of a fibre-forming polymer and an outer sheath of another polymer: improvements are possible with these. Effects like antistatic and antisoil are essentially surface effects; composite fibres should permit changes in surface reactivity and thus improve the efficiency of applied finishes.

A major success with these composite fibres has been achieved in non-woven textiles. To improve the stability and durability of natural non-woven fabrics such as papers requires resin binding of the fibres. With synthetic composite fibres this binding has been achieved by heat-treatment, a mat of fused fibres being obtained. Development leading to new, low-cost apparel and upholstery fabric seems possible. This would suggest that composite fibres may partly supplant the special-effects chemicals. Yet the new non-woven fabrics are prepared by techniques which are relatively undeveloped and it will be surprising if optimization does not create its own demand for some auxiliary chemicals to improve performance of these new processes. For some time to come, therefore, as the synthetic-fibre industry continues to grow, it will demand both more of the existing and also entirely new auxiliary chemicals.

Bibliography

Chapter 1

HAWKINS, W. L., (ed.) (1972). *Polymer stabilization*, Wiley-Interscience, New York.
KUZ'MINSKII, A. S. (1966). *Ageing and stabilization of polymers* (translated by LEYLAND, B. N., 1971), Elsevier, London.
NEIMAN, M. B. (1964). *Ageing and stabilization of polymers* (translated 1965). Consultants Bureau, New York.
SCOTT, G. (1965). *Atmospheric oxidation and antioxidants*, Elsevier, London.

Chapter 2

References to phthalocyanines
BOOTH, G., History and development of phthalocyanine chemistry, *Chimia* (Aarau), 1965, **19**, 201.
MOSER, F. H. and THOMAS, A. L. (1963). *Phthalocyanine compounds*, Reinhold, New York.

Further reading
INMANN, E. R. (1967). *Organic pigments*, Royal Institute of Chemistry Lecture Series, No. 1.
LENOIR, J. (1971). In *The chemistry of synthetic dyes* (vol. 5) (ed. K. Venkataraman), Academic Press, New York.
MASON, S. F. (1970). In *The chemistry of synthetic dyes* (vol. 3) (ed. K. Venkataraman), Academic Press, New York.
Paint technology manuals, part VI (1966). Chapman and Hall, London. Oil and Colour Chemists Association (J. G. Gillan, Collator).
PATTERSON, D. (1967). *Pigments: Introduction to their physical chemistry*, Elsevier, Amsterdam.

Chapter 3

ABRAHART, E. N. (1968). *Dyes and their intermediates*, Pergamon Press, London.
BRADLEY, W. (1958). *Recent progress in the chemistry of dyes and pigments*, Royal Institute of Chemistry, London.
FIERZ-DAVID, H. E. and BLANGRY, L. (1955). *Fundamental processes of dye chemistry*, Interscience, New York.
JUDD, D. B. and WYSZECHI, G. (1963). *Colour in business, science and industry*, Wiley, New York.
SAUNDERS, K. H. (1949). *The aromatic diazo compounds and their technical applications*, Edward Arnold, London.
VENKATARAMAN, K. (1952). *The chemistry of synthetic dyes*, Academic Press, New York.
VICKERSTAFF, T. (1954). *The physical chemistry of dyeing*, Oliver and Boyd, London.
ZOLLINGER, H. (1961). *Azo and diazo chemistry: aliphatic and aromatic compounds*, Interscience, New York.

Chapter 4

GOODMAN, I. (1967). Synthetic fibre forming polymers. Royal Institute of Chemistry Lecture Series No. 3.

GORDON COOK, J. (1968). *Handbook of textile fibres*. Vol. 2. *Man-made fibres*. Merrow Publishing Co. Ltd., Watford.

MARK, H., WOODING, N. S., and ATLAS, S. M. (1971). *Chemical aftertreatment of textiles*, John Wiley, Chichester.

MARSH, J. T. (1966). *Textile science: an introductory manual. An introduction to textile finishing*. Chapman and Hall, London.

MONCRIEFF, R. W. (1970). *Man-made fibres* (5th edn.), Heywood Books, London.

STIMSON, S. C. (1971). *Chem. and Eng. News*, October 4, pp. 24–30.

Index